The Role of Language in the Climate Change

This volume takes a distinctive look at the climate change debate, already widely studied across a number of disciplines, by exploring the myriad linguistic and discursive perspectives and approaches at play in the climate change debate as represented in a variety of genres. The book focuses on key linguistic themes, including linguistic polyphony, lexical choices, metaphors, narration, and framing, and uses examples from diverse forms of media, including scientific documents, policy reports, op-eds, and blogs, to shed light on how information and knowledge on climate change can be represented, disseminated, and interpreted and, in turn, how they can inform further discussion and debate. Featuring contributions from a global team of researchers and drawing on a broad array of linguistic approaches, this collection offers an extensive overview of the role of language in the climate change debate for graduate students, researchers, and scholars in applied linguistics, environmental communication, discourse analysis, political science, climatology, and media studies.

Kjersti Fløttum is professor of French linguistics at the Department of foreign languages, University of Bergen. Her research fields are text and genre theory and discourse analysis, with a special focus on linguistic polyphony in scientific, political, and climate change discourse as well as on narrative structures in climate change discourse. Fløttum has headed several cross-disciplinary projects (KIAP, EURLING, LINGCLIM). She is coauthor of the books *ScaPoLine* (2004) and *Academic Voices* (2006) and editor of *Speaking to Europe* (2013). She has published in international journals such as *Journal of Pragmatics*, *Journal of Language and Politics*, *Language & Communication*, *Text & Talk*, *Climatic Change*, *Global Environmental Change*, *Nature Climate Change* and in *Oxford Research Encyclopedias*. Together with Anwar Saab, "1001Films," Fløttum is co-producer of the documentary film "Talking about climate," available at https://vimeo.com/178449717.

Routledge Research in Language and Communication

1 The Role of Language in the Climate Change Debate
Edited by Kjersti Fløttum

The Role of Language in the Climate Change Debate

Edited by Kjersti Fløttum

Routledge
Taylor & Francis Group
LONDON AND NEW YORK

First published 2017 by Routledge

2 Park Square, Milton Park, Abingdon, Oxfordshire OX14 4RN
52 Vanderbilt Avenue, New York, NY 10017

Routledge is an imprint of the Taylor & Francis Group, an informa business

First issued in paperback 2019

Library of Congress Cataloging-in-Publication Data
A catalog record for this book has been requested.

ISBN: 978-1-138-20959-6 (hbk)
ISBN: 978-0-367-36582-0 (pbk)

Typeset in Sabon
by Apex CoVantage, LLC

Contents

Tables and Figures

Tables

Figures

Foreword

In March 2015, the BBC screened a 90-minute TV documentary titled "Climate Change by Numbers." The program aimed to improve public understanding of climate change by focusing on "just three key numbers that clarify all the important questions about climate change." The three numbers were 0.85 (the degrees Celsius of warming the planet has undergone since 1880), 95 (the percent confidence climate scientists have that at least half this warming is human caused), and 1,000,000,000,000 (tons of carbon it is estimated humans can burn to avoid "dangerous climate change").

Climate change in just three numbers? There is of course the long-standing convention that holds that climate can be defined statistically, as the aggregation and average of (usually) 30 years of standardized meteorological measurements in a place. If one allows for ten weather variables, measured twice daily over 30 years, then the climate of a location would then be described by 219,000 numbers. But that is before one recalculates these numbers over a different period of time to determine how the climate of that place has changed between two periods. So, 438,000 numbers. And that is just the climate of *one* location on a planet with many different climates in different places.

My own early fascination with climate and its changes was triggered by just this love of numbers and the many ways of manipulating them. As a boy I would pore over the daily weather statistics in the local newspaper, and then later, as a university geography student, I would gather together historical weather data from UK meteorological stations to calculate trends in UK winter climate. It was this that placed me on a course of study that was to lead to my career as an academic geographer and professor of climate. And it is through enumerating the physical processes of the Earth system and how these processes can be represented by mathematical equations and solved in computer models that the scientific knowledge assessed by the Intergovernmental Panel on Climate Change (IPCC) has come into being. These IPCC knowledge assessments have informed and been interpreted by government officials and negotiators at successive meetings of the parties to the UNFCCC (United Nations Framework Convention on Climate Change), a process by which the political dimensions of climate change have been identified and through which potential solutions are negotiated.

Enumerated climatic knowledge has certainly come to be regarded by most commentators as the highest and most necessary form of knowledge during the 30 years over which climate change has emerged as a public policy issue. It is this form of knowledge that has inspired some to present climate change as "the defining challenge of our age. The science is clear; climate change is happening, the impact is real. The time to act is now" (Ban Ki-Moon, December 12, 2007, COP13 in Bali).

But there are other languages beyond numbers and mathematics that matter for public debates about climate change. Understanding the public meanings of climate change, and therefore the basis for Ban Ki-Moon's demanded "action," requires *more* than just numbers or scientific knowledge. Studying the ways in which climate change is talked and written about through semantic, visual, and embodied languages, and in different vernaculars, is necessary if the multiple meanings of climate change are to be excavated. And only through the construction and articulation of meaning is personal and collective political action on climate change possible. Science or numbers alone is never enough.

And so the last ten years have seen a growing body of scholarship by linguists and anthropologists that has sought to understand the ways in which climate and its changes are represented beyond numbers. Among professional linguists, Kjersti Fløttum (at the University of Bergen in Norway) and Brigitte Nerlich (at the University of Nottingham in the UK) have led the way. This new book written by Fløttum and her various collaborators—*The Role of Language in the Climate Change Debate*—is therefore a very welcome contribution to this broadening of understanding. Its focus on the linguistic and discursive elements of climate change communication—the importance of frames, narratives, metaphors, rhetoric, and textual analysis methods—reflects this growing international network of analysts. The collection of studies in this book brings us closer to the essentials of linguistic theory and communication practice as they apply to climate change.

To find out what climate *is* (climatology), what it *has been* in the past (paleoclimatology), and what it *might be* in the future (climate prediction) is one thing. But to find out what climate change *means* to different people, in different places and in different times, is something else. And here numbers are not enough. The three numbers in the BBC documentary—0.85, 95, and 1 trillion—are clearly, on their own, inadequate. There is also a need to understand human language(s) and the central role it plays in the formation of public meanings and hence how language shapes the possibilities for political action.

This need is well illustrated by the Climategate controversy that erupted back in the winter of 2009/10. For example, one of the most cited accusations about the integrity of climate scientists was based on the statement in one email from Phil Jones sent to his colleagues in 1999: "I've just completed Mike's [Mann] *Nature* trick of adding in the real temps [temperatures] . . . to hide the decline [in warming]." Was this 'trick' a nefarious practice that subverted the ethics of science (by hiding a cooling), or was it merely a clever statistical device to bring visible order out of complex and chaotic data? Without a proper linguistic analysis of the word 'trick' and how the meaning of this word is constructed by communities of practice—which, after all, is how all language comes to mean anything at all—the ground was ripe for multiple and conflicting interpretations. And this of course is exactly what happened in the subsequent media and public frenzy about the trustworthiness of climate science. Language constructs the spaces within which different forms of political action take place.

The IPCC, through its six yearly assessments of scientific knowledge about climate, has slowly been learning its own lessons about the importance of language. One of the conclusions of the review of processes and procedures of the IPCC, prepared by the Inter-Academy Council in 2010 under commission from UNEP following certain public controversies, was that "communication was a major weakness" in the IPCC. It recommended a professionalization of its communication strategy and more careful consideration of the language used in its reports. And in February 2016, in Oslo, the IPCC held its first expert meeting on climate change and communication, with many recommendations suggesting that "[IPCC] authors should be trained in writing and communicating, including the use of clear language."

As if to press this point home, recent analysis of the language of the IPCC's Summaries for Policy-Makers (SPM) reports has shown their low readability scores in relation not only to mass media but also to other scientific publications (Ralf Barkemeyer and colleagues; *Nature Climate Change* 2015, DOI: 10.1038/NCLIMATE2824). And Kjersti Fløttum, the editor of this new volume, has pioneered other types of linguistic analyses of climate change texts. She has subjected IPCC SPMs and various national climate change policy documents to narrative and polyphonic analysis, to explain how different linguistic and discursive strategies are used to advance the arguments of the scientists or policy officials authoring these important climate change texts. There is no one way of telling the story of climate change, and so being able to discern different voices and communication strategies in key texts such as these is of great public and political value.

Nevertheless, there will remain limits to what can be achieved through the study of climate, its changes, and its public meanings, whether this study be through numbers or through words and images. Such analyses may clarify different forms of knowledge and help identify reasons why specific disagreements and conflicts persist around climate change. And such analyses may explain the effectiveness of different forms of communication—metaphors, visuals, narratives, etc.—to different audiences and thereby inspire new communicative forms. Linguistic analyses also are useful for explaining the particular rhetorical tricks (that word again) that different political actors use on different occasions to persuade particular audiences about desirable futures and courses of action.

Yes, we need to understand the languages of disagreement, of persuasion, and of mobilization. But climate and weather are also experienced bodily, through the senses, through memory, and through livelihoods. These forms of knowing can sometimes exceed the human ability to communicate through words, which helps explain the growing engagement by creative artists—musicians, sculptors, dancers, painters—with the idea of climate change. These different forms of knowing also point to the different meanings, the different structures of reality, carried by words associated with 'weather' and 'climate' in the vernaculars of different regions. Here, the work of anthropologists becomes important.

Climate change is not merely a physical phenomenon; neither is it only a cultural one. But beyond its inescapable material reality, climate change is an idea to which there is no end of meaning and therefore no final resolution to the challenges it presents to people. Although it is an idea that cannot be solved, language remains important to how climate change will be used to change the way people think, plan, argue, act, and prepare for their futures. Living with climate change not only suggests the need for improved *scientific* literacy; it also suggests the need for improved *literacy*—better understanding of how language and communication work in public life, the interplay between language and action in the world. This book, which Fløttum and colleagues have put together, is a good starting point for such inquiry.

Mike Hulme
King's College London
October 2016

Acknowledgements

We are grateful to the Research Council of Norway's SAMKUL program for funding the interdisciplinary LINGCLIM project (grant 220654; www.uib.no/en/project/lingclim), which provided the research environment where most of the studies presented in this volume were conducted. We would like to thank our LINGCLIM colleagues not represented here for fruitful discussions and stimulating collaboration over several years: Gisela Böhm, Anje Müller Gjesdal, Elisabeth Ivarsflaten, and Vegard Rivenes. In addition, Martin Paulsen has helped in systematizing survey data.

We want to thank the NTAP project (Networks of Texts and People; http://ntap.no/) for providing the climate change blog corpus and in particular Dag Elgesem, Knut Hofland, and Samia Touileb.

Special thanks goes to the Norwegian Citizen Panel / DIGSSCORE, which has delivered several of the surveys referred to in this book, especially to scientific director Elisabeth Ivarsflaten, to coordinator Sveinung Arnesen, and to colleagues in the research group "Climate and Environment" for relevant and constructive discussions and collaboration: Endre Tvinnereim, Gisle Andersen, Lise Lund Bjånesøy, Katherine Duarte, Mikael Poul Johannesson, and Åsta Dyrnes Nordø. We extend further thanks to services by *Ideas2Evidence Ltd* (http://www.ideas2evidence.com/en/home).

We are also grateful to the research institute Uni Research, closely affiliated with the University of Bergen, for productive partnership, in particular with the departments Uni Research Rokkan Centre and Endre Tvinnereim, and Uni Research's Center for Big Data Analysis and Andrew Salway.

Special thanks to Professor Mike Hulme for having shown great interest in the LINGCLIM project and for writing the Foreword of this volume.

Finally, in addition to many thanks addressed to our home departments at the University of Bergen (in particular the Department of Foreign Languages), the Norwegian School of Economics, Queen Mary University London, and Oregon State University, we greatly appreciate the interest taken in our research by the Bjerknes Centre for Climate Research.

List of Contributors

Dr. Dimitrinka Atanasova, School of Languages, Linguistics and Film, Queen Mary University London

Professor Trine Dahl, Department of Professional and Intercultural Communication, Norwegian School of Economics

Professor Helge Drange, Geophysical Institute / Bjerknes Centre for Climate Research, University of Bergen

Professor Kjersti Fløttum, Department of Foreign Languages, University of Bergen

Associate Professor Øyvind Gjerstad, Department of Foreign Languages, University of Bergen

Professor Mike Hulme, Department of Geography, King's College London

Assistant Professor Michael D. Jones, School of Public Policy, Oregon State University

Dr. Oleksandr Kapranov, Department of Foreign Languages, University of Bergen

Dr. Nelya Koteyko, School of Languages, Linguistics and Film, Queen Mary University London

Dr. Andrew Salway, Uni Research Computing, Uni Research

1 Language and Climate Change

Kjersti Fløttum

Introduction

In a situation where calls for urgent action on global climate change are heard around the world, is language something that really matters? This rhetorical question constituted an implicit point of departure for the studies presented in this book. Language does of course matter, but the question was, and still is, in what way and to what extent. Current research shows that the meaning people ascribe to climate change (e.g., their understanding of the phenomenon, their perception of the risks involved, the value judgments they make, and the emotional reactions they experience) is closely related to how climate change is portrayed in various contexts of climate communication (Fløttum 2014; Fløttum, Dahl, and Rivenes 2016; Hulme 2009; Leiserowitz 2005; Lorenzoni and Pidgeon 2006; Moser 2010, 2016; Nerlich, Koteyko, and Brown 2010; Pearce *et al.* 2015; Tvinnereim and Fløttum 2015). Language not only reflects and expresses facts and observations; it also influences attitudes and behavior. Language thus constitutes a vital component of the cultural prerequisites underlying societal development, and it is indispensable for interaction and participation—also in the climate change issue. The importance of language is further emphasized through two factors: (1) the complexity of the phenomenon of climate change itself, developed from being understood mainly as physical to becoming political, social, cultural, ethical, and communicational, and (2) the particularly multivoiced and multifaceted nature of the climate change debate, with a wide range of actors and voices, which causes multiple communication challenges due to the high number of stakeholders, interests, opinions, and attitudes represented.

In order to find out more about how language matters in the climate change issue, various kinds of linguistic analyses have been undertaken: studies of words, of combination of words, and of entire texts from different contexts, such as scientific reports, political documents, and media texts (Fløttum 2016). The aim of this book is to present a broad linguistic and discursive view on various questions that arise: How do actors at different levels of decision and in different sectors construct their climate agenda linguistically? How are different voices manifested linguistically? How are arguments and messages organized in entire texts and talks? How is this heterogeneity of information and often value-laden knowledge understood by laypeople? The book aims at answering these questions and showing the crucial role of language in the climate change debate, focusing on both representations and interpretations of climate discourse. Through qualitative as well as quantitative methodological approaches, including a multimodal approach, automatic text analysis, and new data-driven approaches, the eight subsequent chapters cover text material that spans a variety of genres: scientific documents, policy reports and debates, corporate annual reports, newspaper articles, editorials, op-eds, blogs, and 'survey discourse'. The authors take into account different levels of analysis, from a micro-perspective, focusing on particular linguistic phenomena (words, sentences), to a macro-perspective, integrating entire texts and the context they are produced in. Topics that are developed include linguistic polyphony (multivoicedness), epistemic and deontic modality, lexical choices, metaphors, narration (storytelling), and framing. Through genres representing non-expert discourse, there will also be perspectives on attitudes and emotions as manifested in public opinion. The book assumes a global perspective, corresponding to the nature of the climate change issue, through studying empirical material from five continents—Europe, Africa, North America, Asia, and Australia.

The contributing authors are researchers working within, or associated with, the LINGCLIM project ("Linguistic representations of climate change discourse and their individual and collective interpretations," 2013–2016; www.uib.no/en/project/lingclim), based at the University of Bergen, Norway. Under the common overarching research question about the role of language in the climate change debate, new and relevant insights are obtained, contributing substantially to the knowledge base of climate change, in particular to the understanding of public opinion formation. These results constitute much needed input for decision making on what measures to undertake in order to tackle the many challenges of climate change.

There is a growing interest in language related to climate change (Fløttum 2016; Moser 2010, 2016; Nerlich, Koteyko, and Brown 2010), a perspective needed to fill the current knowledge gap on issues related to linguistic representations of climate change knowledge and their interpretations. The present book will address this lack of linguistic and discursive studies and show how climate change text and talk work. The book will thus provide increased awareness of the "human" side of climate change. It aims to highlight the use of language and the development of opinions and attitudes, which will also be of value for current and future decision-making debates. Through increased knowledge and awareness of the role of language, policymakers may be better equipped to deal with the complexities at stake. Explorations of how text and talk about climate change are constructed, circulated, and interpreted may not only add to our knowledge as such but also stimulate increased reflection on relevant concepts and practices.

The book is structured in four parts: "Frames and Narratives" (chapters 2, 3, and 4), "Metaphors" (chapters 5 and 6), "Language of Climate Action" (chapters 7 and 8), and "New Methods to Study the Language of Climate Change" (chapter 9). Short presentations of chapters 2–9 are given below.

Frames and Narratives

In chapter 2, "Verbal and Visual Framing Activity in Climate Change Discourse: A Multimodal Analysis of Media Representations of the IPCC's 5th Assessment Report," Trine Dahl reports on a framing study of media representations related to climate change impacts, adaptation, and vulnerability (see also Dahl 2015). She undertakes a qualitative analysis of news items (text and images) appearing in the two UK newspapers, *The Guardian* and the *Daily Mail*, linked to the recently published IPCC Assessment Report 5. The overarching aim of the study is to explore how multimodal news texts may be analyzed in a framing perspective through combining a linguistic (verbal) and a theme-based (visual) approach. With a point of departure in Entman's (1993) classical definition of framing as relating to selection and salience, she identifies traces of the news producers' framing activity in both the verbal and the visual elements of their texts. The verbal material is discussed mainly in the context of the Appraisal framework developed by Martin and White (2005), with an additional view toward the Discursively Constructed News Values framework outlined in Bednarek and Caple (2012). The visual material is discussed in terms of themes (e.g., O'Neill 2013) along with the visual analysis of news values suggested in Bednarek and Caple (2012). Dahl finds that, in the *Guardian* material, the Settled Science frame (see O'Neill *et al.* 2015) is clearly the most frequently exploited, while no one frame dominates in the *Daily Mail* material. The themes identified in the visual elements in most cases reinforce the framing undertaken through the verbal elements, with the Impacts theme the most frequent by far in both newspapers. In terms of methodological considerations linked to the study of interaction between the two meaning-making resources, Dahl finds that linguistic cues provided through the image captions in many cases play a vital role in contextualizing the visuals, reflecting the challenge of visualizing climate change impacts, many of which have not yet been experienced.

In chapter 3, "Competing Climate Change Narratives: An Analysis of Leader Statements during COP21 in Paris," Øyvind Gjerstad seeks to combine a narrative framework (Fløttum & Gjerstad 2013, 2016) with the analysis of linguistic polyphony or 'multivoicedness' (Nølke, Fløttum and Norén 2004) in official statements made by China, India, South Africa, and the United States on climate change policies during the Conference of the Parties in Paris (COP21). The narrative framework helps map converging and diverging stories, while the polyphonic analysis of linguistic markers such as reported speech, argumentative connectives, and negations has the potential to uncover more or less 'hidden' voices in texts that relate to such a heterogeneous political context (see Fløttum and Gjerstad 2013; Gjerstad 2013). By means of these two approaches, Gjerstad aims to answer the following questions: How do the delegations of the world's two largest emitters and those of newly industrialized nations represent the problem of climate change and its potential solutions in their statements? Who are presented as the main actors (i.e. the heroes, villains, and victims), and, more specifically, what are the roles of various countries in the different narratives? Do the narratives on climate change reflect underlying national and international inequities as well as conflicts regarding historical responsibilities? Gjerstad concludes that the main difference between the four statements is not in policy content but rather in their style. He also finds that the US statement stands out, in that it more clearly resembles a story than the other three. It dramatizes COP21 as a pivotal point in the climate change narrative and paints vivid pictures of possible worlds in the case of both success and failure. As such, the statement appears to be more motivational than policy oriented.

The narrative perspective is taken further in chapter 4, "Stories about Climate Change: The Influence of Language on Norwegian Public Opinion," by Michael D. Jones, Kjersti Fløttum, and Øyvind Gjerstad. Starting with the fact that the general public is exposed to multiple narratives, or 'stories', about climate change, the authors claim that the way these narratives are constructed and linguistically formulated may influence people in various ways (Jones 2014; Jones, Shanahan, and McBeth 2014). The study investigates how differences in the language used in narratives and the structure of narratives may influence people's preferences and perceptions of risk related to climate change. The authors report on an experiment undertaken in March 2015 through a national survey by the Norwegian Citizen Panel, with approximately 335 respondents. The results are analyzed in light of similar experiments undertaken in the United States (Jones 2014). The study is based on two theoretical frameworks: First, the basis for the narrative perspective (inspired by Adam 1992, 2008) relies on the theoretical and methodological text linguistic framework developed by Fløttum and Gjerstad (Fløttum 2013; Fløttum and Gjerstad 2013). Second, the experiment design relies on the Narrative Policy Framework (Jones, Shanahan, and McBeth 2014), which enables an assessment of the role of narrative communication and culture in shaping how information about climate change influences preferences and perceptions of risk. Their findings indicate that the hero character in each story plays a central affective role in determining the persuasiveness of a story.

From the macro-perspective of frames and narratives, the second part of the book moves on to the perspective of words and word combinations creating metaphors in various text genres.

Metaphors

In chapter 5, "Metaphors in Online Editorials and Op-Eds about Climate Change, 2006–2013: A Study of Germany, the United Kingdom, and the United States," Dimitrinka Atanasova and Nelya Koteyko discuss the most frequently used climate-change-related metaphors in 527 editorials and op-eds from three countries that are key to the climate change negotiations (van der Wurff 2009)—Germany, the UK, and the United States. Drawing on the concept of discourse metaphors and using the critical metaphor analysis approach (Charteris-Black 2004), the authors find that the most preferred domain for metaphors is Illness in *Sueddeutsche.de*, War in *theguardian.com*, and Journey in *NYTimes.com*. Atanasova and Koteyko argue that the prominent use of Illness metaphors in *Sueddeutsche.de* opinion-page content, War in *theguardian.com*, and Journey in *NYTimes.com* is an indicator that diverging national efforts to communicate the gravity of anthropogenic climate change are closely intertwined with national contexts and cultures, political identities, and policy making. Such tailoring to national identity and culture can be seen as a positive step to maximize public resonance by appealing to a set of recognizable values and norms. But there also is a potential danger in this prominent use of those metaphors, as Illness, War, and Journey are deeply entrenched in political communication. The authors suggest that the transfer of these metaphors to climate change discourse risks turning the issue into "politics as usual"—a development that defies advocacy for new and creative metaphors as a way to provoke climate change action.

Metaphors are also the object of study in chapter 6, "Conceptual Metaphors Associated with Climate Change in Corporate Reports in the Fossil Fuels Market: Two Perspectives from the United States and Australia," where Oleksandr Kapranov undertakes a qualitative analysis of conceptual metaphors associated with the issue of global climate change identified in annual reports by ExxonMobil (US) and BHP Billiton (Australia). The material of the study involves these corporations' official 2014–2015 annual reports, available to the public online. The analysis involves identification of conceptual metaphors viewed through the lenses of cognitive linguistics. Within this framework, conceptual metaphor is regarded as a robust means of discursive space construction. The qualitative study further presented in this chapter seeks to juxtapose conceptual metaphors associated with climate change in corporate discourse by ExxonMobil and BHP Billiton, respectively. The data analysis reveals that the conceptual metaphors involved in the framing of climate change discourse by BHP and ExxonMobil is represented by several identical conceptual metaphors—e.g., 'Climate Change as a Challenge', 'Climate Change as a Risk', 'Climate Change as a Low-Carbon Future', and 'BHP/Exxon as a Climate Change Aware and Responsible Corporation'. However, ExxonMobil's discursive space is also characterized by the corporation-specific conceptual metaphors 'Climate Change as an Energy-Efficient Future' and 'Climate Change as Natural Gas Future'. These findings are suggestive of two approaches adopted by these corporations. Specifically, ExxonMobil construes its climate change discourse via conceptual metaphors associated with natural gas and efficient energy consumption, and BHP frames its climate change discourse by means of conceptual metaphors evocative of climate change mitigation as a process, a path, which requires knowledge and investment.

After investigating how climate change is represented in various genres and from various perspectives in parts one and two, part three of the book focuses on how actions to tackle climate change are viewed both by the general public and by politicians.

Language of Climate Action

In chapter 7, "Willingness of Action," Kjersti Fløttum takes as a point of departure the multiple voices expressing diverse points of view according to interests, values, and worldviews in the public debate on climate change. From being to a large extent focused on the causes of climate change (the attribution issue), the debate has now become much more oriented toward what measures to undertake to meet the many predicted challenges at different levels (local, national, and global). In order to obtain new knowledge about the opinions and attitudes of the public in this matter, the following open-ended survey question has been asked through the online survey Norwegian Citizen Panel: "Concerning climate change, what do you think should be done?" In addition to a content (topic) analysis (Roberts *et al.* 2014), this chapter undertakes in-depth linguistic analyses of the 4,634 respondents' answers to this question, with a particular focus on deontic perspectives (how the world ought to be, what actions are needed, and whose responsibility they are) and polyphonic (multivoiced) expressions. The purpose is to uncover how the public debate has influenced people's engagement in questions related to the transition to a low-carbon society and their willingness to act (Kotchen *et al.* 2013; Obradovich and Guenther 2016; Tvinnereim and Ivarsflaten 2016). The study provides new knowledge on some of the fundamental constraints on and opportunities for climate action, which are essential to decision making.

In chapter 8, "The Paris COP21 Agreement—Obligations for 195 Countries," Kjersti Fløttum and Helge Drange take a closer look at what is requested by the Paris Agreement from December 2015. In the aftermath of this Paris COP21 agreement, where voices applauding the "historical" event are paralleled by voices pointing to the necessity of immediate implementation (Clémençon 2016; Savaresi 2016), some crucial questions may be raised: How are scientific findings, which the requested actions are expected to be based on, presented? And what is the rhetorical tone of the document? This chapter investigates these two questions in a linguistic analysis of the science-policy interface, moving from the epistemic perspective ('policy neutral') present in expressions of probability in the most recent Synthesis Report (2014) published by the IPCC to the deontic perspective present in expressions of policy obligations in the COP21 agreement from the UN climate summit in Paris. The analysis shows that the rhetorical tone reflects the necessity of political action, based on recent scientific conclusions, but in a sufficiently general way to obtain the necessary approval by the various national assemblies.

New Methods to Study the Language of Climate Change

In chapter 9, "Data-Driven Approaches to Climate Change Discourse, Illustrated through Case Studies of Blogs and International Climate Negotiation", Andrew Salway undertakes a review of automated methods to investigate language use in the climate change debate. He argues that the scale of climate change discourse necessitates the incorporation of automatic text analysis techniques into research methods. Furthermore, he argues that the complexity and variety of relevant textual material mean that data-driven approaches in particular have an important role in order to minimize prior assumptions about the content and form of the material to be analyzed. After considering what data-driven means in the context of automated text analysis and reviewing commonly used techniques, this chapter presents three case studies and reflects on the data-driven text analysis techniques that were incorporated into corpus-based methods for investigating climate change discourse. The main issue is the extent to which automated techniques help to elucidate linguistic patterning that maps to relevant phenomena—e.g., topics, representations, and frames. In the first case study, a word list, sorted concordances, and simple templates are used to induce nine meaning categories that characterize representations of the future in a corpus of approximately 1.5 million blog posts; these categories are the basis for a large-scale comparison of how the concept of the future is treated in "accepting" and "skeptical" blog communities. The other two case studies use a novel technique—local grammar induction—in order to identify salient information structures in the same blog corpus and in a corpus of international climate change negotiations. Structures induced from the blog corpus seem to highlight characteristics of individual blogs. Structures induced from climate change negotiations are used to elucidate the positions of different countries and relations between them.

To sum up—the studies presented in this book aim to show the potential of language at both word and text levels for representation and interpretation purposes of the climate change phenomenon. Since climate change—in contrast to the weather and the consequences of climate change—cannot be seen, heard, or touched, it is in fact through language that we can acquire knowledge of and understand this complex phenomenon. What is not articulated (orally or in writing) can be a basis neither for decisions nor for actions. Increased knowledge on the use of language will thus contribute to better decision making about different pathways of action for tackling climate change.

References

Adam, J.-M. 1992. *Les textes: Types et prototypes*. Paris: Nathan.

Adam, J.-M. 2008. *La linguistique textuelle. Introduction à l'analyse textuelle des discours*. Paris: Armand Colin.

Bednarek, M., and H. Caple. 2012. "'Value added': Language, image and news values." *Discourse, Context & Media* 1: 103–113.

Charteris-Black, J. 2004. *Corpus approaches to critical metaphor analysis*. Basingstoke & New York: Palgrave MacMillan.

Clémençon, R. 2016. "The two sides of the Paris climate agreement: Dismal failure or historic breakthrough?" *Journal of Environment & Development* 25 (1): 3–24. doi: 10.1177/1070496516631362

Dahl, T. 2015. "Contested science in the media: Linguistic traces of news writers' framing activity." *Written Communication* 32 (1): 39–65.

Entman, R. M. 1993. "Framing: Toward clarification of a fractured paradigm." *Journal of Communication* 57 (1): 51–58.

Fløttum, K. 2013. "Narratives in reports about climate change." In: Gotti, M., and C. S. Guinda (eds.). *Narratives in academic and professional genres*, pp. 277–292. Bern: Peter Lang.

Fløttum, K. 2014. "Linguistic mediation of climate change discourse." *Asp—la revue du GERAS* 65: 7–20.

Fløttum, K. 2016. "Linguistic analysis approaches for assessing climate change communication." In: Von Storch, H. (ed.). *Climate science: Oxford research encyclopedias*. Oxford: Oxford University Press. doi: 10.1093/acrefore/9780190228620.013.488

Fløttum, K., T. Dahl, and V. Rivenes. 2016. "Young Norwegians and their views on climate change and the future: Findings from a climate concerned and oil rich nation." *Journal of Youth Studies* 19 (8): 1128–1143. http://dx.doi.org/10.1080/13676261.2016.1145633

Fløttum, K., and Ø. Gjerstad. 2013. "Arguing for climate policy through the linguistic construction of narratives and voices: The case of the South-African green paper 'National Climate Change Response'." *Climatic Change* 118 (2): 417–430. doi: 10.1007/s10584-012-0654-7

Fløttum, K., and Ø. Gjerstad. 2016. "Narratives in climate change discourse." *WIREs Climate Change*. doi: 10.1002/wcc.429

Gjerstad, Ø. 2013. "Voices and identities: The polyphony of Tony Blair and Nicolas Sarkozy the European Parliament." In: Fløttum, K. (ed.). *Speaking of Europe: Approaches to complexity in European political discourse*, pp. 111–134. Amsterdam: John Benjamins Publishing Company.

Hulme, M. 2009. *Why we disagree about climate change*. Cambridge: Cambridge University Press.

Jones, M. D. 2014. "Cultural characters and climate change: How heroes shape our perceptions of climate science." *Social Science Quarterly* 95 (1): 1–39.

Jones, M. D., E. A. Shanahan, and M. K. McBeth. (eds.) 2014. *The Science of stories: Applications of the narrative policy framework in public policy analysis*. Basingstoke & New York: Palgrave MacMillan.

Kotchen, M. J., K. J. Boyle, and A. A. Leiserowitz. 2013. "Willngness-to-pay and policy- instrument choice for climate-change policy in the United States." *Energy Policy* 55: 617–625.

Leiserowitz, A. 2005. "American risk perceptions: Is climate change dangerous?" *Risk Analysis* 25: 1433–1442.

Lorenzoni, I., and N. Pidgeon. 2006. "Public views on climate change: European and USA perspectives." *Climatic Change* 77: 73–95.

Martin, J., and P. White. 2005. *The language of evaluation: Appraisal in English.* Basingstoke & New York: Palgrave MacMillan.

Moser, S. C. 2010. "Communicating climate change: History, challenges, process and future directions." *WIREs Climate Change* 1: 31–53.

Moser, S. C. 2016. "Reflections on climate change communication research and practice in the second decade of the 21st century: What more is there to say?" *WIREs Climate Change* 7 (3): 345–369. doi: 10.1002/wcc.403

Nerlich, B., N. Koteyko, and B. Brown. 2010. "Theory and language of climate change communication." *WIREs Climate Change* 1: 97–110.

Nølke, H., K. Fløttum, and C. Norén. 2004. *ScaPoLine: La théorie scandinave de la Polyphonie linguistique.* Paris: Kimé.

Obradovich, N., and S. Guenther. 2016. "Collective responsibility amplifies mitigation behaviors." *Climatic Change* 137: 307–319.

O'Neill, S. J. 2013. "Image matters: Climate change imagery in US, UK and Australian newspapers." *Geoforum* 49: 10–19.

O'Neill, S. J., H. T. P. Williams, T. Kurz, B. Wiersma and M. Boykoff. 2015. "Dominant frames in legacy and social media coverage of the IPCC Fifth Assessment Report." *Nature Climate Change* 5 (April): 380–385.

Pearce, W., B. Brown, B. Nerlich, and N. Koteyko. 2015. "Communicating climate change: Conduits, content, and consensus." *WIREs Climate Change* 6 (6): 613–626. doi: 10.1002/wcc.366

Roberts, M. E., B. M. Stewart, D. Tingley, C. Lucas, J. Leder-Luis, S. K. Gadarian, et al. 2014. "Structural topic models for open-ended survey responses." *American Journal of Political Science* 58: 1064–1082.

Savaresi, A. 2016. "The Paris agreement: A new beginning?" *Journal of Energy & Natural Resources Law* 34 (1): 16–26. doi: 10.1080/02646811.2016.1133983

Tvinnereim, E., and K. Fløttum. 2015. "Explaining topic prevalence in answers to open-ended survey questions about climate change." *Nature Climate Change* 5: 744–747. doi: 10.1038/nclimate2663

Tvinnereim, E., and E. Ivarsflaten. 2016. "Fossil fuels, employment, and support for climate policies." *Energy Policy* 96: 364–371. doi: http://dx.doi.org/10.1016/j.enpol.2016.05.052

van der Wurff, R. 2009. "Climate change policy of Germany, UK and USA." In: Boersema, J. J., and L. Reijnders (eds.). *Principles of Environmental Sciences*, pp. 459–471. New York: Springer.

Frames and Narratives

2 Verbal and Visual Framing Activity in Climate Change Discourse

A Multimodal Analysis of Media Representations of the IPCC's 5th Assessment Report

Trine Dahl

Introduction

The Intergovernmental Panel on Climate Change (IPCC) is the world's most influential climate organization, evaluating the science related to climate change. Their comprehensive Assessment Reports (AR)—published every five to six years since 1990—represent the outcome of a painstaking process involving three Working Groups (WG), dealing with the physical science basis (WG I), impacts (WG II), and solutions (WG III). The last decade or so has seen the public debate on climate change broadening to encompass aspects that go far beyond the purely physical one (e.g., Hulme 2009). This is reflected in media accounts, where climate 'stories' may span fields ranging from economics to psychology and art. While it is commonly accepted in media and communication research today that news reports are never truly objective or neutral (Cotter 2010; White 2012), news consumers do not necessarily notice that a story is given a specific framing, or angle. To bring this out, communication scholar Robert M. Entman (1991, 6) suggests:

> Comparing media narratives of events that could have been reported similarly helps to reveal the critical textual choices that framed the story but would otherwise remain submerged in an undifferentiated text. Unless narratives are compared, frames are difficult to detect fully and reliably, because many of the framing devices can appear as 'natural,' unremarkable choices of words or images.

The notion of framing concerns how individuals and groups organize, perceive, and communicate about the world. Framing studies are common within many social science disciplines (de Vreese 2005; Entman 1993), and many of them take their point of departure in Entman's (1993, 52) definition:

> To frame is to select some aspects of a perceived reality and make them more salient in a communicating text, in such a way as to promote a

> particular problem definition, causal interpretation, moral evaluation, and/or treatment recommendation for the item described.

The key words in this definition are *selection* and *salience*. However, different disciplinary traditions have different conceptualizations of framing and different methodological approaches to how the notion can be operationalized. If we look at this from the news producers'[1] perspective, framing can be considered a process that implies a (conscious or subconscious) strategic selection of verbal, and often also visual, features in order to make certain aspects of an issue or event salient to the intended readership. Using Entman's (1993) terminology, this implies combining the *text* and *communicator* perspective of framing,[2] with certain language—and image—elements seen as representing strategic choices made by the communicator in order to promote a particular 'take' on the issue/event.

Framing studies involving climate discourse have been undertaken by means of various, often quantitative, methods and with various aims. Commonly seen approaches in the literature are content analysis studies (e.g., identifying what frames are available to / used by a society at a given time [e.g., O'Neill *et al.* 2015]), surveys (e.g., revealing how climate change is perceived by specific audiences [e.g., Corner *et al.* 2011]), and psychological experiments (e.g., studying how influential a particular frame is in terms of spurring people to take action [e.g., Weber and Stern 2011]). The aims and methodological approach of the study presented here are different. I undertake a qualitative analysis of multimodal (language/image) news items dealing with the IPCC AR5 WG II report entitled "Climate Change 2014: Impacts, Adaptation, and Vulnerability,"[3] in order to investigate how news writers in two national newspapers in the UK verbally and visually framed their stories relating to this report (see next section for research questions).The publication of this IPCC report had been anticipated with some apprehension. The WG I report (2013; on the physical science basis) had sent the very clear message that "[w]arming of the climate system is unequivocal, and since the 1950s, many of the observed changes are unprecedented over decades to millennia."[4] It therefore seemed obvious that both present and future impacts on humans and societies in many parts of the world would be outlined as very serious, indeed. The key communicative device chosen by many scientists to convey WG II's findings was risk language (Painter 2015a). It was hoped that presenting impacts, as well as solutions (WG III report), as a risk issue would make the information more accessible and engaging to non-experts. Painter (2015a) discusses whether this strategy actually worked, concluding that, for journalists at least, they "did not pick up on the language of risk in a volume and detail commensurate with the degree to which it was being heavily sponsored by some IPCC co-chairs and lead authors" (287). Painter's explanations for this include the reflection that "the concept of 'risk management' sounds like a specialist or jargon-ridden term, which most journalists find problematic" (287) and that a "disaster frame lends itself to a strong narrative and pictures . . . , whereas explicit risk is more of an issue or concept than a story" (287).

In the next sections, I start by briefly stating my aims and research questions, before I describe the material. Then, I present the method used to analyze the verbal and the visual representations, respectively. Next follows a fairly comprehensive case study involving four of the news texts, in order to demonstrate the analytical procedure in some detail, before I report and discuss my findings for the whole material. Finally, some concluding remarks are offered.

Aims and Research Questions

The overarching aim of the study is to explore how multimodal news texts may be analyzed in a framing perspective through combining a linguistic (verbal) and theme-based (visual) approach. Through an investigation of how news writers in two important UK news providers (*The Guardian* and the *Daily Mail*; see the next section for details) framed the reported information from the AR5 WG II report for their readership, the following three research questions are addressed: (i) How can the notion of framing in verbal representations be operationalized within a linguistics-based framework? (ii) What frames were exploited in the two news sources? (iii) And how does the verbal and visual mode of communication operate together in the context of framing?

Material Analyzed

As already indicated, the texts I analyze in this chapter come from two UK mainstream media sources (online versions): *The Guardian* (theguardian.com) and the *Daily Mail* (dailymail.co.uk). *The Guardian* (GUA) is a left-leaning quality newspaper that over a long period has dedicated substantial resources to reporting on climate change and environmental issues (O'Neill *et al.* 2015; Painter 2015b) from a diversity of angles. The *Daily Mail* (DM) is a conservative middle-market tabloid that has often provided climate skeptics with a platform for advocating their mistrust in mainstream climate science in general and in the IPCC in particular (Boykoff and Mansfield 2008; Dahl and Fløttum Forthcoming; Jaspal, Nerlich, and Koteyko 2013). It might therefore be hypothesized that the framings chosen by the news producers in the two sources are likely to diverge. The texts for the present study were identified through the search function of each newspaper's website, with the search items 'IPCC' and 'climate change' in addition to each of the following keywords from the WG II report: 'impact', 'adaptation', and 'vulnerability'. These searches yielded 24 texts from GUA and 13 from DM. As this is a multimodal study, all the texts returned through the keyword searches were checked for visual elements (pictures or graphics). It turned out that at least one picture was included in every text. The

GUA texts typically had one visual, but three texts had three or four. The total number of visuals for the GUA material was 35 (32 pictures and three graphical elements). As for the DM texts, the typical text had one or two visual elements, but one text had six, and another had as many as seven. The total number of visuals for the DM material was 39 (32 pictures and seven graphical elements). In terms of genre, the GUA material comprises news reports, opinion/comment pieces, and blog contributions, while the DM material primarily consists of news reports and a few opinion/comment pieces. The texts cover the time span March 22–April 3, 2014, for GUA, and March 18–April 5, 2014, for DM.

Method for Analyzing Framing Activity in Multimodal Media Texts

As indicated in the "Introduction," framing is a commonly used concept within social science research. This has, as a consequence, led to a framing literature that displays a broad selection of frame indicators (the features/elements said to contribute to establishing a frame) as well as frame labels (e.g., Dahl 2015; de Vreese 2005; O'Neill *et al.* 2015). Typically, each new study employs a unique frame set and to some extent also basis for establishing a frame, something that makes it difficult to compare findings across studies.

In the wake of the publication of the IPCC AR5 in 2013–2014, O'Neill *et al.* (2015) undertook a comparison of how the report was framed in mainstream media (broadcast and print) and on social media (Twitter) in the UK and United States. The ten frames they established were based on certain "constituent elements" (2015, 380 and Supplementary Information;[5] see also below). Descriptions of all the frames are available from http://www.nature.com/nclimate/journal/v5/n4/box/nclimate2535_BX1.html. In the following listing, short descriptions are given for those labels that may not be so intuitively understood: 'Settled Science' (SS; emphasis on the science of climate change and the broad expert consensus; considerable evidence of the need for action), 'Political or Ideological Struggle' (PIS; a conflict over the way the world should work, over solutions or strategy to address climate change), 'Role of Science' (ROS; explores the role science plays in society; may debate transparency, funding, or public awareness, especially in relation to institutions—e.g., the IPCC), 'Uncertain Science' (US), 'Disaster' (D), 'Security' (S; threat to human security; could be energy, water, or food security, or a threat to a nation state—e.g., migration), 'Morality and Ethics' (ME), 'Opportunity' (O), 'Economic' (E), and 'Health' (H).

O'Neill *et al.*'s (2015) discourse-based approach rooted in social science considered categories like narratives, themes, and stock phrases/metaphors, in addition to descriptions of visual imagery. This chapter, in contrast, takes its point of departure in the toolkit of linguistics, or—more specifically—what may be termed 'text linguistics', for the discussion of the verbal part of the material (see below for details), focusing on micro-level features (see further below). The visual elements are considered primarily in terms of theme (e.g., O'Neill 2013). Many studies of visuals are theme based (see the next section for further examples). In the current context, theme is understood as the visual equivalent of the pretheoretical notion of topic, what a text is about (Brown and Yule 1993). In this sense, the theme represents an indication of the angle, or frame, selected by the news producer. To enable synergies with the O'Neill *et al.* (2015) study, I will use their frame labels as a point of departure for the analysis. In addition, in the study of the interaction between the verbal and visual mode of communication, I also look to Bednarek and Caple's (2012) discursive news values analysis, developed to analyze multimodal news texts.

Analytical Approach: The Verbal Material

In my analysis of news producers' framing activity in the verbal part of their texts, I consider how linguistic resources are used for framing purposes. First, lexical elements are identified that indicate which domain (the term is here used in a somewhat loose sense) the writer has chosen to place the issue/phenomenon in. Examples of domains may be science, politics, economics—or security, catastrophe/disaster, scientific uncertainty, depending on the frame set considered for a particular study (see above). In addition, the concept of evaluation (e.g., Hunston and Thompson 2000) is seen as playing an important part in the framing process. Within text linguistic analysis, evaluation is an overarching notion that concerns the interpersonal perspective of texts (e.g., Hyland 2005). I here make use of a relatively simple version of the complex and multilayered Appraisal framework (Martin and White 2005; White 2012) to deal with micro-level features considered to convey evaluation. In the Appraisal framework, attitudes and value positions are seen as reflected in the linguistic choices made by the writer. This kind of micro-level analysis involves careful reading of individual texts. Evaluative meaning is not necessarily embodied in a specific linguistic element but often the product of the linguistic co-text or broader context.[6] Some words, like *success* and *failure*, may be described as inherently positive and negative, respectively; but consider the word *negative*. In the statement *They were negative to the proposal*, the word is used in its basic, negative, meaning. However, in a statement like the following, uttered in a medical context, the same word contributes to a positive interpretation: *The test results came back negative*. In addition, specific combinations of words may result in a negative or positive interpretation, sometimes depending on the reader's own worldview (e.g., White 2012). To present and discuss the outcome of such a fine-grained qualitative approach for complete texts would be very space consuming. I will therefore focus on the headline, which has been shown to have a strong framing effect on news stories (e.g., Andrew 2007; Bell 1991; Molek-Kozakovska 2016).

As already mentioned, a framework that has been developed to analyze both verbal text and pictures is Bednarek and Caple's (2012) discursively constructed news values, or discursive news values analysis (DNVA; M. Bednarek, personal communication, November 19, 2014). The Bednarek and Caple (2012) study shows—through examples from a case study based on reports from an Australian broadsheet newspaper on the 2011 Queensland floods—how news events are constructed as newsworthy. In this framework, evaluation and attitude are key notions in the analysis of news values constructed through language. This makes it methodologically feasible to combine framing analysis based on the Appraisal framework with discursive news values analysis.

Analytical Approach: The Visual Material

Analyzing pictures may be said to be more challenging than analyzing verbal text. First of all, a picture lacks devices to indicate temporal, spatial, and logical connections (O'Neill and Smith 2014). Perhaps more importantly, pictures have both denotative (literal, non-associative) and connotative (associative) attributes (Wozniak, Lück, and Wessler 2015), and the latter will to some extent vary from person to person (e.g., Nicholson-Cole 2005). Studies of climate-change-related visuals have often involved theme-based classifications. These classifications are characterized by the same label diversity as those seen in framing studies (see above), due to the bottom-up approach typically applied in order to account for the specific material discussed in each study. For instance, DiFrancesco and Young (2011)—in a rare combined verbal/visual study—establish 15 climate change themes for the visuals: e.g., Political, Scientist/expert, Urban landscape, Polar bear, and Transportation. Smith and Joffe (2009) identify three broad themes: Impact, Personification, and Graphical representations. And O'Neill (2013) identifies the themes of People, Impacts, Protest, Causes, Solutions, Sci & Tech, Weather, and Other. A different angle is offered by Nerlich and Jaspal (2014), who undertake a thematic analysis of emotional responses to climate imagery related to extreme weather events. Their analysis establishes four themes for their material, with—apart from the first theme—negative emotional meanings: *people in the developing world 'getting on with it'*; *threat and danger from extreme weather*; *premature ageing and deterioration of the earth*; and *human-induced extreme weather*. In the discussion of the visual elements that are part of my material, I will look to these four studies but mainly draw on the themes established in O'Neill (2013).

In the DNVA framework, the visual analysis comprises the contextualization of image participants (where/who?) as well as photographic devices including, e.g., angle, which may be linked to attitude and stance (Bednarek and Caple 2012, 105–106). In their case study texts, the news values of Negativity, Impact, and Superlativeness (the maximized or intensified aspect of an event) were commonly seen in both modes of communication. In the following headline from my material, *Climate change: The poor **will** suffer most* (*The Guardian*, March 31), Negativity and Impact are simultaneously construed through the linguistic choices made by the writer. The picture included in the text shows the complete destruction of a village, with inhabitants and relief workers starting on the daunting task of clearing the area of debris from crushed houses and everything else that used to make up this community. The caption tells us that the picture is taken in the aftermath of the typhoon and storm that hit Tacloban (the Philippines) in November 2013. The news values represented in the visual element are again Negativity and Impact, but also very much Superlativeness, through the depiction of total devastation. As in Bednarek and Caple (2012), these three news values were very common in my material, but some headlines and pictures drew on other values, as can be seen in the 'Main Study' section below.

The Verbal and Visual Framing of the AR5 WG II Report

As stated above, my first research question asks how the notion of framing may be operationalized within a linguistics-based framework. Since a textual presentation and discussion of the kind of qualitative analysis attempted here is very space consuming, a total of 37 texts (even when considering the headline only) would be too much to present in any great detail. I will therefore illustrate the analytical procedure through a case study of four texts, before presenting findings from the complete material. The four texts were all published on March 31, 2014, and are written by experienced environment/science journalists. The texts form two pairs that both deal with roughly the same issue. The first pair, GUA 1 and DM 1 (see next section for details), focuses on the threat climate change poses to mankind, while the second, GUA 2 and DM 2, establishes links between the global and the national perspective.

Case Study

The four news items that have been singled out to demonstrate in some detail how we may study framing activity at the interface of text (cf. Entman's *text* location) and text producer (the *communicator* location) are as follows: GUA 1: "Climate change a threat to security, food and humankind—IPCC report" (by Suzanne Goldenberg); DM 1: " 'No one will be untouched': Climate change will lead to war, famine and extreme weather, claims IPCC report" (by Ellie Zolfagharifard); GUA 2: "Climate catastrophes in other countries will hit UK's food security, experts say" (by John Vidal); and DM 2: "UK warned of 'climate change flood of refugees': Droughts and heatwaves could force millions to flee their country" (by Ben Spencer). Through the linguistic analysis of the headlines, I will try to identify framing elements pointing to a particular (primary) frame[7] selected by the news producer, as well as the discursively constructed news values drawn on to make the text newsworthy to readers. The visual elements are analyzed in terms of main theme (indicating a frame) and news value(s).

Headlines, Linguistic Analysis

The headline of GUA 1 clearly fits O'Neill *et al.*'s (2015) Security (S) frame description, linguistically signaled as such through the phrase "threat to security, food and humankind," indicating global validity of the statement. The DM 1 headline, while also referring to food security, plays more directly on the Disaster (D) frame through the journalistic value of personalization, expressed through the individually angled *No one . . . untouched* and *famine* (in contrast to GUA 1's all-encompassing and impersonal *humankind*, and the 'objectified' *food*, with less expressed immediate relevance to individuals than *famine*).

The DM 1 writer makes categorical, high-probability statements regarding predicted consequences: "No one **will be** untouched" and "Climate change **will lead to** war, famine and extreme weather"—the latter headline comprising the inherently negative words *war* and *famine* and the contextually negative *extreme weather*. Both the GUA 1 and DM 1 headlines include a so-called attribution phrase. The attributed source of the statements in both instances is the IPCC report. The attribution phrase of GUA 1 contains the reporting verb *report*, while the attribution phrase in DM 1 is introduced by the verb *claim*. In English-language news texts, reporting verbs like *say*, *state*, and *report* are by far the most common (e.g., Cotter 2010; Dahl and Fløttum 2014; White 2012). These verbs are neutral in terms of the writer's stance toward the reported statement, while other verbs may indicate attitudinal value (endorsement or distancing). As for *claim*, selected by the headline producer of DM 1, its attitudinal value is not clear cut but may in context be seen as conveying distancing from, rather than endorsement of, a statement. In the current case, such an interpretation would be in line with DM's long-time tradition of providing a platform for voices portraying the IPCC as a scaremongering institution, exaggerating the climate change threat (e.g., Dahl and Fløttum Forthcoming).

When we turn to the second pair of texts, GUA 2 and DM 2, we see that again the headline of the GUA text explicitly frames it as a security issue, this time with a national focus and related to food only (*will hit UK's food security*). The DM text, too, is framed as a national security issue but here related to migration (*UK warned of 'climate change flood of refugees'*). Both articles are similar to GUA 1 and fit O'Neill *et al.*'s (2015) description of the S frame (see above). In this text pair, we see that it is the GUA headline that makes a categorical, high-probability claim ("**will** hit"), while DM 2 expresses a less categorical, medium-probability claim ("**could** force").

In DM 2, so-called graduation elements (Martin and White 2005) increase the force of the statements: "**flood of** refugees," "**millions** [of people]." Such linguistic upscaling is also quite prevalent in the body of both DM articles.

Warnings are *dire*, change, declines, and impacts will be *dramatic*, conflicts *violent*, infrastructure damage *serious*, nations *swamped* with *a tide of* refugees, while, on the positive side, the understanding of what needs to be done has improved *dramatically*, and opportunities for adapting are *huge*.

Headlines, News Values Analysis

In terms of the discursively constructed news values in the headlines, Negativity and Impact are both embodied in all four texts, mainly through the verbal element (GUA 1: ***a threat to***; DM 1: ***will lead to war***; GUA 2: ***catastrophes will hit***; DM 2: ***warned of . . . droughts*** *and* ***heatwaves could force*** *millions*). The two DM headlines in addition exploit the news value of Superlativeness (DM 1: ***No one*** *will be untouched*; DM 2: ***flood*** *of refugees*; *force* **millions** *to*). The two GUA headlines both draw on the news value of Eliteness[8] (GUA 1: ***IPCC*** *report*; GUA 2: ***experts*** *say*). DM 1, too, mentions the IPCC, but the potential distancing expressed through the reporting verb *claims* weakens the Eliteness value potential of the name and hence also the trustworthiness and force of the message as such.

Visuals, Theme Analysis

Finally, let us turn to the visuals of the four texts. As mentioned in the "Introduction," Painter (2015a) finds that a "disaster frame lends itself to a strong narrative and pictures" (287). The DM 1 headline was found to set up a Disaster frame for the text through personalization (see above). The only picture in this text is a screen shot of a female news reporter on MSNBC. This is a realization of the People theme, but clearly not related to the affected people referred to in the headline of this text. The other visual elements of the text consist of six graphical IPCC representations of climate change phenomena—e.g., world maps and charts of projected temperature rises and decreases in crop yields. Such graphical elements were found to be typical of the D frame in O'Neill *et al.* (2015, Supplementary Information). The graphical elements thus make connections to *extreme weather* and *famine* in the headline. The color coding related to projected temperature changes represents the strongest and clearest visual effect (moving from pale yellow to deep red and purple; cf. Mahony and Hulme 2012).

The Security-framed GUA 1 text, on the other hand, includes a picture of huts and trees appearing as small islands in a sea of dirty grey water, and doll-like people standing in small groups outside their homes or trying to move between the 'islands'. The caption reads: "Flooding in Pakistan in 2010, the kind of extreme weather events which the IPCC says is the result of climate change." The exploited theme in the image is clearly Impacts. Flooding-related images make up the largest category in Nerlich and Jaspal's (2014) study. They suggest that, in their material, the human impact images from a non-Western context tend to convey associations of people 'getting on with it', despite the serious effects of the flooding, which is portrayed as a distant phenomenon. However, the GUA 1 picture rather seems to convey the paralyzing feeling of 'What do we do now?' in the face of a disaster. However, the aerial photo does not show any close-ups of human faces, and hence no emotional signals can be read from it. The links to the verbal elements of the headline (*threat to . . . food and humankind*) are easy to make, but the picture personalizes the more abstract account in the headline, pulling the framing from Security toward Disaster.

In DM 2, three pictures are included. The first shows a group of male asylum seekers moving along a rail track, which the caption tells us is in Calais, France. Some of the men walk with bowed heads or faces covered by scarves, while others display tentative and perhaps hopeful smiles. The main theme may be said to be People. The second picture is an iconic image of cracked earth, again realizing the Impacts theme (cf. Nerlich and Jaspal's 2014 *premature ageing and deterioration of the earth*), with a caption linking it to global warming and its broader consequences. Impacts is also the theme of the third picture, showing a flooded housing estate/village, with no people in sight. The caption locates it to the Songkhla Province, Thailand, 2005. Two of the three visuals of this article, then, link up with the headline (*drought* and *flood of refugees*), even though it is not obvious that the depicted group of asylum seekers are migrating for climate-change-related reasons.

Finally, GUA 2 provides an Impacts theme picture of a flooded Western countryside, displaying bales of hay floating in the water, with no humans or human activity visible. The caption locates the scene to Yorkshire, England. In this case, the article headline focuses on *climate catastrophes in other countries*, while the visual element in fact depicts consequences of a local/national weather incident.

Visuals, News Values Analysis

When it comes to news values expressed through these visuals, Negativity and Impact are again key values simultaneously construed in most of them. The exception is the TV presenter image in DM 1, which may be interpreted as realizing the news value of Eliteness. The DNVA framework is limited to photographs (Bednarek and Caple 2012). However, similar to the extensive flooding pictures from Pakistan (GUA 1) and Thailand (DM 2), the deep red and purple color of some of the IPCC graphical illustrations of DM 1 might be seen as embodying the news value of Superlativeness, in addition to Negativity and Impact.

Relationship Verbal—Visual: Frame Harmony or Dissonance?

The third research question of this study asks how the verbal and visual modes of communication operate together in the context of framing. A news report exploiting different angles in verbal text and visual(s) runs the risk of presenting an unclear message to the readers. The analysis of the four case study texts resulted in an overall impression of frame harmony between the two meaning-making resources. We see this clearly in text DM 1. Through the linguistic analysis of the headline, it was placed in the Disaster frame category, due to the personalized angle expressed through the headline words *no one* and *famine*. The graphical elements of the text convey the serious consequences of temperature rises, corroborating the verbal 'story'. The three other texts—GUA 1, GUA 2, and DM 2—were all classified as exploiting the Security frame in terms of headline wording. As for the visuals of these texts, GUA 2 (Impacts: flooded hay field) and DM 2 (Impacts: drought; People: migrants; Impacts: flooded village) had pictures that fit well with the S frame. In GUA 1, on the other hand, the Impacts picture of people maneuvering in a flooded environment triggered associations to the Disaster frame. However, the S and D frames appear as closely related and compatible, and their combination in this news text does not result in frame dissonance. O'Neill *et al.* (2015) provide the following comparison between D and S: "[C]rop failure is D, but S would detail impacts on food supply." The only additional potential dissonance between the verbal and the visual was observed in GUA 2, where the headline introduced the issue as distant with national implications (*Climate catastrophes* ***in other countries*** *will hit UK's food security*), while the visual portrayed it as a local problem (a flooding picture from Yorkshire, UK).

As for news values, Negativity and Impact were identified as the main values exploited in both the verbal and the visual part of all four texts. Comparing the two newspapers, we see that GUA and DM differ in the former's exploitation of the Eliteness value linked to the attributed source (*IPCC*) in the two headlines, while the latter either did not mention the sources behind the WG II report (DM 2) or potentially questioned their expert status (DM 1, ***claims*** *IPCC report*).

Main Study

Having demonstrated in some detail how a linguistic/thematic/news value analysis of framing activity in multimodal texts may be undertaken, I will now present findings from the analysis of all the 37 texts. One of the GUA texts is entitled *Climate change 'felt on all continents and across the oceans'* (March 28), echoing case study text DM 1, and thus indicating the pervasiveness of climate change impacts (with drought or flooding and ensuing crop failure as important aspects). However, when all the texts from these two UK news sources are considered, it is clear that much of the coverage focused on distant areas, in both the verbal and the visual representations. This may be a reflection of the fact that the WG II report makes clear that the vulnerability aspect is much more serious for regions and populations far away from the UK. In terms of frames exploited by the news producers, the body part of the individual texts often contained traces of several of the frames established in O'Neill *et al.* (2015). The linguistic analysis undertaken in the present study, however, focuses on the overall article framing undertaken through the headline. Findings for the 37 headlines are presented in Table 2.1. A set of five frames (listed alphabetically) is posited as sufficient to account for the analyzed texts: Disaster (D), Opportunity (O), Role of Science (ROS), Security (S), and Settled Science (SS).

Table 2.1 Main Frame Exploited, Headlines

FRAME	*GUA (n = 24)*	*DM (n = 13)*
Disaster	5	2
Opportunity	-	1
Role of Science	1	3
Security	2	5
Settled Science	16	2

Table 2.1 reveals that a clear majority of the GUA texts were framed as SS (16 out of 24). The choice of this frame implies that the news writer presents the WG II report's claims as valid and that they should be acted upon. In the DM material, the S frame was the most common of the frames represented (five out 13 texts). This frame, too, indicates belief in the IPCC message, but its inherent conflict potential (e.g., related to food shortages and migration) perhaps makes it a better 'fit' for a tabloid in the journalistic quest for newsworthiness (Cotter 2010, 68–69). We notice that the DM material also comprises one text where the journalist exploits the Opportunity frame, expressed through inherently positive vocabulary: *Climate change will* ***benefit*** *tourism and cause* ***a boom*** *in Arctic cruises, claims leaked IPCC report* (March 28, 2014).

None of the headlines explicitly addressed the WG II report's key focus on adaptation, apart from general references to *action*, which may potentially refer to both adaptation and mitigation (cutting CO_2 emissions). A few of the DM texts in the body part included comments related to the construction of flood defenses and storm shelters; a quoted source stated that, due to food scarcity, *we might have to change our diets*; another stated, *Ultimately we, may be looking at scenarios where we have to help resettle climate refugees*; and a very specific suggestion was made as part of a list of warming consequences leading to changed snow patterns: *Downhill ski resorts could use snow machines*. In the GUA material, somewhat less concrete suggestions include the proposal to *improve the economy's resilience to the likely impacts of climate change*, and that *[i]mmediate and transformative action is needed at every level: individual, local, national; personal, political, and financial*. In a contributed text by a government advisor on climate change, Lord Deben, the following question is raised: *Why go to all the trouble of curbing emissions when we can just build higher sea walls or grow different crops?* It is answered: *This is a seductive view, but ultimately, a naively mistaken one*.

As for the visual elements (pictures only), Table 2.2 presents the main themes observed for the two newspapers.

Table 2.2 Main Theme, Visuals (Pictures Only)

THEME	*GUA (n = 32)*	*DM (n = 32)*
Causes	4	-
Impacts	17	20
People	5	7
Protest	1	1
Weather	-	1
Other	5	3

O'Neill (2013), in her study of climate change reporting in three countries, found that pictures portraying climate adaption and mitigation were rare. This was also the case in the present material, despite the focus on adaptation in the WG II report. The climate impacts theme was the most important one by far in both newspapers (GUA: 17; DM: 20). This is a theme that clearly presents more opportunities for visualization than adaptation and that has been shown to be highly salient to news consumers (O'Neill *et al.* 2013). A few of the GUA pictures classified as Impacts could have been placed in the People category instead. These pictures showed clearly recognizable distressed individuals, but the Impacts theme was deemed to be the more striking of the two alternatives. Such analytical decision making will also be influenced by the connotative attributes the analyst assigns to the image. These may be triggered by personal values and experiences as well as by the verbal context of the visual. In the present material, the associative leaps between the picture and the headline or caption were in some cases substantial (see below). A final thing to note is that while four of the 32 GUA images represented a climate change Causes theme (e.g., an oil-drilling scene), none of the 32 DM images did.

As for the news values construed through language and image, Table 2.3 provides an overview of the material from both newspapers. Note that each

Table 2.3 Main News Values, Headlines and Pictures

NEWS VALUE	*GUA*		*DM*	
	Headlines (n = 24)	*Pictures (n = 32)*	*Headlines (n = 13)*	*Pictures (n = 32)*
Eliteness	16 (67%)	3 (9%)	10 (77%)	5 (16%)
Impact	12 (50%)	9 (28%)	11 (85%)	20 (63 %)
Negativity	15 (63%)	16 (50%)	10 (77%)	23 (72%)
Superlativeness	6 (25%)	3 (9%)	7 (54%)	4 (13%)
Unclear news value	-	11 (34%)	-	-

Each headline/picture may embody more than one news value.

headline/picture may embody several news values. Only the values deemed most prominent have been recorded here. Percentages are included to bring out trends in the material, but the number of texts analyzed is clearly far too small to make generalizations about multimodal climate news discourse on this basis.

Unsurprisingly, the Eliteness value is present in many headlines of both newspapers, through the mention of *IPCC*. Considering headline and picture together, Negativity and Impact are the two most important values. Notably, the DM texts frequently draw on these two news values. Other noteworthy trends are the percentage of DM headlines exploiting the Superlativeness value, and the 'Unclear news value' of roughly a third of the pictures in GUA (e.g., a pair of dark hands around coffee beans and a single bird flying past Big Ben).

Another news value in the DNVA framework is Aesthetics. Interestingly, several pictures, notably in the GUA material, were very aesthetically pleasing, contradicting the grave message of the text as such. Numbers for this value have not been included here, because other news values were considered more salient. The Aesthetics value was even more prominent in media reports of the WG I report on the physical science, with, e.g., accounts of global warming and retracting glaciers accompanied by pictures of beautiful pearl-like drops of water on clear blue backgrounds (Dahl and Fløttum Forthcoming).

So, in terms of the interplay between the two modes of communication, did the material discussed in this chapter display frame harmony or dissonance? In the majority of the texts analyzed, the verbal and visual reinforced each other's message. However, in a few cases, this relationship may be said to be one of dissonance, with the potential to send a confusing message to the readers. The combination that seemed most odd was seen in a DM article entitled *QUANTUM LEAP: India's climate disaster in the making* (April 1). The only visual accompanying the text was a picture of several cars covered in a thick layer of snow in what seems to be a wintry Northern European city, reflected in a visible building and the clothing of a man walking through the snow. The picture caption tells us that *[w]ith several parts of the country experiencing extreme weather events, India could soon face several climatic catastrophes*. However, the image as such depicts a Weather rather than Impacts theme; in the context of Northern Europe, the signification is neither extreme weather nor disaster/catastrophe.

Concluding Remarks

This chapter has explored how news producers' framing activity in multimodal climate 'stories' may be analyzed. With a basis in the news reports by *The Guardian* and the *Daily Mail* on the IPCC AR5 part-report "Climate change 2014: Impacts, Adaptation, and Vulnerability," the verbal representations are analyzed in terms of linguistic features linked to selected domain and to evaluation, while the visuals are analyzed in terms of theme. In addition, exploited news values have been identified in both language and image representations.

In terms of specific findings, the analysis showed that both newspapers made use of several of the frames established in O'Neill *et al.* (2015), with 'Settled Science' (implying calls for action) as the dominant frame in the material from the *Guardian* (16 of 24 headlines), and the conflict-related 'Security' frame (threats to nations) as the most common one (five of 13 headlines) in the tabloid *Daily Mail*. The themes identified in the visuals accompanying the texts in most cases reinforced the framing undertaken through the headlines, with the Impacts theme the most frequent by far in both news sources. In many cases, the verbal cues provided through the image captions played a vital role in contextualizing the visuals. One such example was the Calais asylum seeker image accompanying the case study text DM 2 (*UK warned of 'climate change flood of refugees'*), which could have served several other news contexts and framings as well. The caption is required to activate the link to the headline: "Britain and other Western nations could face a tide of refugees as the Earth warms, a new report warns. Pictured, asylum seekers walking along a railway line in Calais, France." Another such example was a typical tourist brochure picture in the DM material of a beautiful Caribbean beach: white sands, blue-green water, palm trees, and a sailing boat, with the caption "Caribbean beaches and other small islands which tend to be shallow could disappear by rising sea levels." Finally, the interaction between the two modes of communication in terms of these elements has been considered. Not surprisingly, the news values analysis of both headlines and images confirmed Bednarek and Caple's (2012) findings that Negativity and Impact are key values in reports on climate change. The headlines of the tabloid *Daily Mail* in addition frequently drew on the news value of Superlativeness through linguistic upscaling (***flood** of refugees*, ***violent** conflict*).

An important aspect of this study has been to assess the usefulness of specific analytical tools for dealing with the concept of framing in multimodal news texts. The text linguistic approach—based on lexical indications of domain in addition to evaluation (Martin and White 2005; White 2012)—used to analyze the verbal representations made it possible to bring out "critical textual choices that framed the story" (Entman 1991, 6). The theme-based analysis employed to determine the topic of the visual part of the 'story' used the categories established in O'Neill (2013) and was able to account for most of the material (with eight of a total of 64 pictures in the non-specific 'Other' category). However, in several cases, a picture fit more than one theme category, and a (subjective) choice had to be made; in addition, as already indicted, the caption was often needed to contextualize the image. As for the news values framework developed by Bednarek and Caple (2012), this was found to be well suited to deal with the verbal text.

As regards the visual news values analysis, the main focus of the DNVA framework is image participants and emotions, and the pictures analyzed in their case study were all linked to a specific event (flooding incident). In the current study, no such specific event was involved; rather, the news texts were triggered by the IPCC report. The visual material did comprise depictions of people with visible facial expressions; however, most of the pictures were not of this kind. In addition, almost half of them were generic ones from image banks. These are decontextualized representations, not linked to specific people, places, or events (e.g., Machin 2004). Even though generic images are ubiquitous in the press today, their frequent use in climate 'stories' may also be seen as reflecting the specific challenge of visualizing climate change impacts and adaptation measures, many of which have not yet been experienced or developed.

Notes

1 The terms *news producer*, *news writer*, and *journalist* are all used in this chapter. The reason for this variation is that many of them are journalists, but some are invited contributors with other backgrounds. In addition, newsroom practices imply that headlines are generally produced by an editor/sub-editor, while visual elements may be selected by section editors or dedicated picture editors.
2 The other two perspectives, or what Entman calls "locations in the communication process" (1993, 52–53), are *receiver* and *culture*.
3 See http://www.ipcc.ch/report/ar5/wg2/.
4 See https://www.ipcc.ch/pdf/assessment-report/ar5/wg1/WGIAR5_SPM_brochure_en.pdf.
5 See http://climatesouthwest.brightcraft.co.uk/library/ONeill_et_al_2014_Dominant_frames_media_IPCC_Supp_Info.pdf.
6 Computer-assisted analysis of, for instance, so-called concordances may also be used to uncover the function of individual words in context (see, e.g., Bednarek and Caple 2012, 2014; Potts *et al.* 2015): A list of occurrences of a particular word and its immediate linguistic surroundings (co-text) are produced, usually in the so-called KWIC (keyword in context) format, revealing patterns in the company the word keeps.
7 Notice that one text may draw on several frames (cf., e.g., Dahl 2015; Porter and Hulme 2013).
8 In Bednarek and Caple (2012), this news value is termed Prominence, while Eliteness is used in Bednarek and Caple (2014). The latter label seems most appropriate for most of the present instances.

References

Andrew, B. C. 2007. "Media-generated shortcuts: Do newspaper headlines present another roadblock for low-information rationality?" *The International Journal of Press/Politics* 12 (2): 24–43.

Bednarek, M., and H. Caple. 2012. "'Value added': Language, image and news values." *Discourse, Context & Media* 1: 103–113.

Bednarek, M., and H. Caple. 2014. "Why do news values matter? Towards a new methodological framework for analyzing news discourse in critical discourse analysis and beyond." *Discourse & Society* 25 (2): 135–158.

Bell, A. 1991. *The language of news media*. Oxford: Blackwell.
Boykoff, M., and M. Mansfield. 2008. "Ye Olde hot air: Reporting on human contributions to climate change in the UK tabloid press." *Environmental Research Letters* 3 (2): 1–8.
Brown, G., and G. Yule. 1993. *Discourse analysis*. Cambridge: Cambridge University Press.
Corner, A., D. Venables, A. Spence, W. Poortinga, C. Demski, and N. Pidgeon. 2011. "Nuclear power, climate change and energy security: Exploring British public attitudes." *Energy Policy* 39: 4823–4833.
Cotter, C. 2010. *News talk: Investigating the language of journalism*. Cambridge: Cambridge University Press.
Dahl, T. 2015. "Contested science in the media: Linguistic traces of news writers' framing activity." *Written Communication* 32 (1): 39–65.
Dahl, T., and K. Fløttum. 2014. "A linguistic framework for studying voices and positions in media texts on climate change." *Text and Talk* 34 (4): 401–420.
Dahl, T., and K. Fløttum. Forthcoming. "Reporting on the future: A news values-based analysis of language and image in newspaper texts on climate change".
de Vreese, C. H. 2005. "News framing: Theory and typology." *Information Design Journal and Document Design* 13 (1): 51–62.
DiFrancesco, D. A., and N. Young. 2011. "Seeing climate change: The visual construction of global warming in Canadian print media." *Cultural Geographies* 18: 517–536.
Entman, R. M. 1991. "Framing U.S. coverage of international news: Contrasts in narratives of the KAL and Iran air incidents." *Journal of Communication* 41 (4): 6–27.
Entman, R. M. 1993. "Framing: Toward clarification of a fractured paradigm." *Journal of Communication* 57 (1): 51–58.
Hulme, M. 2009. *Why we disagree about climate change: Understanding controversy, inaction and opportunity*. Cambridge: Cambridge University Press.
Hunston, S., and G. Thompson (eds.). 2000. *Evaluation in text: Authorial stance and the construction of discourse*. Oxford: Oxford University Press.
Hyland, K. 2005. "Stance and engagement: A model of interaction in academic discourse." *Discourse Studies* 7 (2): 173–192.
Jaspal, R., B. Nerlich, and N. Koteyko. 2013. "Contesting science by appealing to its norms: Readers discuss climate science in the Daily Mail." *Science Communication* 35: 383–410.
Machin, D. 2004. "Building the world's visual language: The increasing global importance of image banks in corporate media." *Visual Communication* 3 (3): 316–336.
Mahony, M., and M. Hulme. 2012. "The colour of risk: An exploration of the IPCC's 'Burning Embers' diagram." *Spontaneous Generations: A Journal for the History and Philosophy of Science* 6 (1): 75–89.
Martin, J., and P. White. 2005. *The language of evaluation: Appraisal in English*. Basingstoke & New York: Palgrave MacMillan.
Molek-Kozakowska, K. 2016. "Stylistic analysis of headlines in science journalism: A case study of *New Scientist*." *Public Understanding of Science*. doi: 10.1177/0963662516637321
Nerlich, B., and R. Jaspal. 2014. "Images of extreme weather: Symbolising human responses to climate change." *Science as Culture* 23 (2): 253–276.
Nicholson-Cole, S. A. 2005. "Representing climate change futures: A critique on the use of images for visual communication." *Computers, Environment and Urban Systems* 29: 255–273.

O'Neill, S. J. 2013. "Image matters: Climate change imagery in US, UK and Australian newspapers." *Geoforum* 49: 10–19.
O'Neill, S. J., M. Boykoff, S. Niemeyer, and S. A. Day. 2013. "On the use of imagery for climate change engagement." *Global Environmental Change* 23: 413–421.
O'Neill, S. J., and N. Smith. 2014. "Climate change and visual imagery." *WIREs Climate Change* 5: 73–87.
O'Neill, S. J., H. T. P. Williams, T. Kurz, B. Wiersma, and M. Boykoff. 2015. "Dominant frames in legacy and social media coverage of the IPCC Fifth Assessment Report." *Nature Climate Change* 5 (April): 380–385.
Painter, J. 2015a. "Taking a bet on risk." *Commentary, Nature Climate Change* 5 (April): 286–288.
Painter, J., 2015b. *The effectiveness of the IPCC communication: a survey of (mainly) UK-based users*. Report written as part of the Joint Programming Initiative (JPI) AR5 in Europe project.
Porter, K. E., and M. Hulme. 2013. "The emergence of the geoengineering debate in the UK print media: A frame analysis." *The Geographical Journal* 179 (4): 342–355.
Potts, A., M. Bednarek, and H. Caple. 2015. "How can computer-based methods help researchers to investigate news values in large datasets? A corpus linguistic study of the construction of newsworthiness in the reporting of Hurricane Katrina." *Discourse & Communication* 9 (2): 149–172.
Smith, N. W., and H. Joffe. 2009. "Climate change in the British press: The role of the visual." *Journal of Risk Research* 12 (5): 647–663.
Weber, E., and P. C. Stern. 2011. "Public understanding of climate change in the United States." *American Psychologist* 66 (4): 315–328.
White, P. R. R. 2012. "Exploring the axiological workings of 'reporter voice' news stories–Attribution and attitudinal positioning." *Discourse, Context & Media* 1: 57–67.
Wozniak, A., J. Lück, and H. Wessler. 2015. "Frames, stories and images: The advantages of a multimodal approach in comparative media content research on climate change." *Environmental Communication* 9 (4): 469–490.

3 Competing Climate Change Narratives

An Analysis of Leader Statements during COP21 in Paris

Øyvind Gjerstad

Introduction

The 21st Conference of the Parties in late 2015 (COP21) was widely regarded as a pivotal moment in the international effort to curb climate change (CC) and deal with its consequences. After previous failures to reach a binding agreement, pressure was on the leaders and negotiators convening in Paris (see also chapter 8 in this volume). On November 30, at the start of the conference, 150 leaders of state and government convened at the Leaders Event to mark the start of the negotiations (UNFCCC 2015). Among them were the leaders of China, India, South Africa, and the United States. All four countries are significant emitters of greenhouse gases (GHG), but their historical responsibilities for emissions and levels of economic development are vastly different. The Chinese economy has undergone profound changes in terms of industrialization, and the country has overtaken the United States as the world's biggest polluter. India and South Africa are both emerging economies with a need for sustained economic growth in order to fight poverty. On the international stage, South Africa has taken the lead among developing nations by heading the Group of 77 and China, having also hosted the COP17 in Durban in 2011. As the biggest historical emitter and second-largest current emitter of GHG, the United States has recently vowed to take the lead in the fight against CC, notably in the "U.S.-China Joint Presidential Statement on Climate Change," published in late 2014, one year before COP21. Along with a number of other countries, the leaders of China, India, South Africa, and the United States offered statements to outline their respective positions and set the tone for the negotiations. These speeches thus had considerable strategic significance, which leads to the following questions: How do the delegations of the world's two largest emitters and those of newly industrialized nations represent the problem of climate change and its potential solutions in their statements? In their narratives about climate change, who are presented as the main actors (i.e., the heroes, villains, and victims), and, more specifically, what are the roles of various countries according to the different statements? Do the narratives on CC reflect underlying national and international inequities, as well as conflicts regarding historical responsibilities? The answers to these questions can bring to light the strategic positioning of the four countries at the start of the negotiations at COP21. I will use two main theoretical tools in my analysis: (1) Fløttum and Gjerstad's approach to CC narratives (2013a, 2013b), and (2) the Scandinavian theory of linguistic polyphony, or *ScaPoLine* (Nølke, Fløttum, and Norén 2004, see also chapter 7 in this volume). According to the narrative approach, text and talk about climate change tend to have a narrative structure, a plot leading from an initial situation—through a complication, a reaction, and a resolution—to a final situation (Fløttum and Gjerstad 2013a, 2013b, 2016). If we consider the statements through the prism of this five-part structure, it is possible to account for similarities and differences between the texts in the various stages of the stories on climate change. The ScaPoLine will be used to complement the narrative analysis, through the identification of voices other than that of the speaker at the moment of utterance, which are brought about through different linguistic markers, such as reported speech or negation. Through the analysis of such markers, we can paint a picture not only of how the speaker views the debate around a given topic but of how the different voices form part of the rhetorical strategy, by appearing in alignment with or in opposition to the speaker, be it implicitly or explicitly.

This chapter will be structured as follows: The following part presents and explains the theoretical framework, before the analysis, which is organized according to the narrative structure, beginning with the initial situation and ending with the final situation. The chapter closes with concluding remarks.

Analytical Tools: Narrative Theory and Linguistic Polyphony

The analysis is based on two unrelated but complementary theories. The approach to climate change narratives developed by Fløttum and Gjerstad is rooted in text linguistics—more precisely, in the work of J.-M. Adam (1992, 2008)—while the polyphonic analysis studies linguistic markers such as words and sentence structure and is based on the Scandinavian theory of linguistic polyphony, or ScaPoLine (Nølke, Fløttum, and Norén 2004), which is rooted in French enunciative linguistics.[1] The narrative approach looks at the text as a whole or at the level of sequences, or paragraphs, while the polyphonic analysis complements this macro-perspective by considering relevant phenomena at the micro-level of words and sentences.

According to Adam (2008), the prototypical narrative sequence is made up of five components, such as in the following example:

> (1) (**Initial situation**) Three young friends went swimming in the Annecy lake. (**Complication**) One of them, Janinetti, went missing. (**Reaction**)

The others dove in. (**Resolution**) They managed to pull her out, (**Final situation**) but she was already dead.

The plot is represented by the second, third, and four components, which constitute the heart of the narrative. Thus, the stability of the initial situation is disrupted by the complication. The reaction seeks to reestablish the equilibrium, leading to a resolution. The final situation represents a new state of stability—in this case the outcome is negative.

For Adam, the narrative sequence is one of five prototypical text sequences, the others being the descriptive, argumentative, explicative, and dialogical sequences (1992). His primary concern is to identify the criteria that can help to separate the narrative sequence from the others, leading him to concentrate on storytelling. However, Fløttum (2010) has proposed that the five narrative stages can be used to describe not only stories but also discursive representations of climate change, what she calls *climate change narratives*. Such a climate change narrative could for example be centered on the COP15 negotiations in Copenhagen in 2009, as follows (Fløttum and Gjerstad 2013a: 77–78):

(2) **Initial situation:** CO_2 emissions increased dramatically between 1990 and 2007. **Complication:** Global warming has caused serious problems in numerous regions. **Reaction:** The UN organized an international summit in Copenhagen in 2009 (COP15) to discuss action on climate change. **Resolution:** But the negotiating countries did not reach any binding agreement of measures to undertake. **Final situation:** Climate change constitutes a serious threat to the planet, and those who have contributed least to the problems are the ones most vulnerable to the consequences.

We see that the issue of climate change easily gives rise to a narrative plot, due to the dramatic nature of the issue. This has also been observed in studies of policy texts such as government white papers (Fløttum and Gjerstad 2013a, 2013b) and reports published by the International Panel on Climate Change (Fløttum 2013), where several of Adam's narrative components have been identified. This approach has also been inspired by the Narrative Policy Framework (NPF), which studies policy narratives from the perspective of political science (see Jones, McBeth, and Shanahan. 2014). One important contribution of the NPF has been to show the centrality of characters—i.e., heroes, villains, and victims—in policy narratives (Jones 2014).

The following example from Chinese President Xi Jinping's remarks at the Leaders Conference can serve to illustrate the relevance of the narrative analysis regarding the issue of climate change, which constitutes the primary complication of the story, much in the same way as in example (2):

(3) (REACTION 1) Thanks to joint efforts of all parties since the United Nations Framework Convention on Climate Change entered into force

> over 20 years ago, global actions on climate change have made progress (COMPLICATION 2) although there are still numerous difficulties and challenges. (REACTION 2) This Paris Conference is hence convened to strengthen implementation of the UNFCCC and bring about a comprehensive, balanced, ambitious and binding agreement on climate change.

In this story, the primary reaction corresponds to the international process of the UNFCCC, which thus far has proven insufficient, leading to a secondary complication. COP21 is then put forth as a secondary reaction, with the aim of arriving at a positive resolution and final situation. As we will see in the analysis, this basic narrative structure is shared by all four statements.

In the present chapter, the narrative approach is mainly applied to the macro-level of larger paragraphs. It constitutes a type of content analysis that can be applied across disciplines, as it does not require linguistic expertise.[2] The linguistic component of the analysis is thus restricted to the polyphonic markers which are found in the examples.

According to the ScaPoLine (Nølke, Fløttum, and Norén 2004), there are linguistic phenomena that signal the presence of other voices than that of the speaker at the moment of utterance, giving rise to 'multivoicedness', or polyphony (see also Gjerstad 2013). Quotations and other types of reported speech represent the most evident cases of this, such as the following example taken from President Obama's statement, in which he makes use of the moral authority of Dr. Martin Luther King to argue his point (emphasis mine):

> (4) For I believe, **in the words of Dr. Martin Luther King, Jr.**, that there is such a thing as being too late.

Other linguistic markers create a more subtle interplay between the voices, such as the use of negation by President Xi Jinping:

> (5) The Paris Conference is **not** the finishing line but a new starting point.

The negation is polyphonic because it constructs a point of view (POV) that is refuted by the speaker:

POV 1: "The Paris Conference is the finishing line"
POV 2: "POV 1 is erroneous"

The subtlety of this polyphony is related to the fact that there is no explicit voice behind this refuted POV, nor is there any claim on the part of the speaker that the opinion is held by anyone, which can be exploited by the speaker for strategic purposes. There are many other markers of polyphony, such as argumentative connectives (e.g., *therefore*; see also example 6), and markers of epistemic modality (*maybe*), by which the speaker can express uncertainty regarding the veracity of a given POV.

Furthermore, example (5) can be used to show the relevance of polyphonic theory for narrative analysis. In fact, in a narrative perspective, the interaction between different points of view may constitute not only singular cases of agreement or disagreement but meeting points between complete storylines. In the context of climate change narratives, the negation in (5) is used not only to refute a POV but to counter an opposing narrative, one where the story ends in Paris. By doing so, President Xi seeks to replace this narrative with his own.

Analysis

The conference in Paris opened in a spirit of collaboration, China having published joint statements with the United States in November 2014 and India in May 2015. The similarities of narratives constructed in the statements made during the Leaders Event on November 30 reflect this consensus. As the analysis will show, there are significant differences between the four narratives, both in content and in style. However, they all fit into a common structure, an overarching narrative, illustrated in Figure 3.1. As we will see in the following, this shared structure can serve as a framework for the comparative analysis, which will be structured according to the temporal succession of the different phases.

The common narrative structure outlined in Figure 3.1 echoes the broad international consensus on the causes and consequences of climate change and the political process to deal with the problem. The continued increase in GHG emissions after previous international negotiations constitutes a secondary complication. The corresponding secondary reaction is a fork in the road, followed by either success or failure. Not all of the components

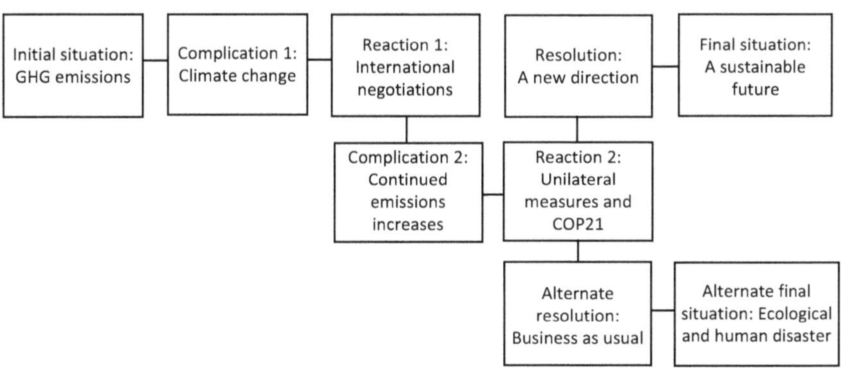

Figure 3.1 Common Narrative Structure

appear in all four statements. Many elements remain tacit: Both historical emissions, climate change, previous negotiations, and continuing emissions serve mainly as a contextual backdrop for present unilateral measures and the summit in Paris—i.e., the second reaction of the narrative. However, to the extent that the earlier stages of the narrative are mentioned, they are not devoid of strategic significance. This is particularly so in the case of equity and historical responsibilities of developed nations, which are discussed by all four leaders, in ways that reflect their countries' respective positions and interests. In the following, we will look at each narrative component as constructed by the four statements.

Initial Situation and Complication 1: GHG Emissions and Climate Change

Not surprisingly, the initial situation and first complication do not feature prominently in the statements, as the origins and effects of GHG emissions are considered to be well-established facts. Nevertheless, the disparity in historical emissions does appear as an argument in all four texts, and it is presented as a principle guiding the negotiations (words and syntactic constructions that produce polyphonic interpretations are marked in bold):

> (6) China: **Given** the difference between developed and developing countries in historical responsibility, developing stage and coping capability, the principle of common but differentiated responsibilities, **instead of** being obsolete, must continue to be adhered to.

Through the use of *given*, example (6) starts by presupposing a number of differences between developed and developing countries, among them historical responsibility. In other words, the existence of these differences is presented as the point of view of a common voice, exempted from debate. Furthermore, this presupposition serves as an argument for a conclusion, which is that 'the principle of common but differentiated responsibilities must continue to be adhered to', in opposition to the point of view that 'the principle of common but differentiated responsibilities are obsolete'. The voice behind this point of view is not identified, but developed countries would be a likely candidate, corresponding to a major difference in the CC negotiation narrative. Marked by *given*, the example as a whole is constructed as an enthymeme, or reduced syllogism, in which the major premise remains implicit:

> **Minor premise**: There is a difference between developed and developing countries in historical responsibility, developing stage, and coping capability.
> **Major premise**: If there is a difference between developed and developing countries in historical responsibility, developing stage, and coping

capability, then the principle of common but differentiated responsibilities, instead of being obsolete, must continue to be adhered to.

Conclusion: The principle of common but differentiated responsibilities, instead of being obsolete, must continue to be adhered to.

The major premise is attributed to a collective voice and ensures the logical validity of the relationship between the minor premise and the conclusion. Its rhetorical force derives from the fact that it is not explicitly mentioned but reconstructed mentally by the recipients, who therefore become 'complicit' in its construction, even if they do not agree with the truth, validity, or relevance of it. The line of argument in example (6) would seem to correspond to the strategic interests of China, which has resisted calls for changes to its status as a developing country in light of strong economic growth in recent decades.

In his statement, Indian Prime Minister Narendra Modi also constructs the question of historical responsibility as a settled issue:

> (7) India: We hope advanced nations will assume ambitious targets and pursue them sincerely. It is **not** just a question of historical responsibility. They also have the most room to make the cuts and make the strongest impact.

The negation serves to refute an implicit point of view according to which "it is just a question of historical responsibility," and that may be attributed to developed nations. By denying the exclusiveness of this issue, the statement also contributes to cementing it as a given principle.

The argument also appears in the South African President Jacob Zuma's remarks:

> (8) South Africa: Developed countries, that have the greatest historical responsibility, must honour their existing commitments and continue to take the lead to address climate change. We are here today largely due to industrial policy choices that now threaten development.

While using historical responsibility to argue the same point as China and India, the South African statement is the only one that elaborates on the meaning of the concept, a fact that reflects its centrality in the global CC narrative.

The US statement is the only one that describes the first complication, while not mentioning the term *historical responsibility:*

> (9) USA: This summer, I saw the effects of climate change firsthand in our northernmost state, Alaska, where the sea is already swallowing villages and eroding shorelines; where permafrost thaws and the tundra burns; where glaciers are melting at a pace unprecedented in modern times. . . .
>
> I've come here personally, as the leader of the world's largest economy and the second largest emitter, to say that the United States of America

> not only recognizes our role in creating this problem, we embrace our responsibility to do something about it. . . .
>
> **We know the truth that** many nations have contributed little to climate change but will be the first to feel its most destructive effects.

Through the description of climate change in the United States, the statement constructs the country as a victim, a characteristic that is later modified as the President evokes his position as leader and representative, thereby establishing himself as the voice of his nation, to officially concede to its role as historical villain. The promise of future action then serves to pivot the country toward a hero role. The portrayal of the United States as both victim and villain on the issue of historical responsibility stands in contrast to the singular villain role attributed to developed countries by the Chinese, Indian, and South African statements, at this stage of the narrative. The last segment of example (9) underscores the injustice of climate change through the formulation "we know the truth that," thereby constructing this point of view as belonging to a collective voice—i.e., as a matter of universal agreement.

Reaction 1: International Negotiations

In his remarks, President Obama did not speak of previous international efforts to curb GHG emissions, but the three others, and most notably President Zuma, did:

> (10) China: Thanks to joint efforts of all parties since the United Nations Framework Convention on Climate Change entered into force over 20 years ago, global actions on climate change have made progress **although** there are still numerous difficulties and challenges.

Through the use of the concessive connective *although* in example (10), President Xi balances two points of view, corresponding respectively to the first reaction and the secondary complication of the narrative (see Figure 3.1). All members of the UNFCCC are here cast as heroes.

> (11) India: [Climate justice] also means aggressive mitigation action by developed countries before 2020, including ratification of 2nd commitment period of Kyoto Protocol, removing conditions and revisiting targets.
>
> (12) South Africa: For South Africa, the outcome of the Paris conference has significant meaning. As Leaders, we are today because of the historic and bold decisions we took in Durban to enhance the implementation of the Convention. You would recall that before Durban, this multilateral process on climate change was under serious threat. South Africa felt duty bound to ensure that it emerged from Durban intact to give hope to humanity. For South Africa, the Durban decision to enhance

the implementation of the Convention was of paramount importance, because the impacts of climate change are harshest on the poor.

The South African statement is the only one that elaborates on previous negotiation efforts, in effect constructing a micro-narrative of the Durban negotiations. In example (12), Zuma paints a dramatic picture of "a serious threat" that is averted by South Africa as the hero committed to protect the poor, who are presented as the victims. In its portrayal of these characters, the example echoes the *South African Climate Change Response White Paper*, which was published in 2011 ahead of the negotiations in Durban (see Fløttum & Gjerstad 2013b).

Complication 2: Continued Emissions Increases

Only two of the statements touch on continued emissions as the result of inadequate efforts to curb emissions:

> (13) China: [T]here are still numerous difficulties and challenges.

> (14) South Africa: Climate change impacts are driven by global inaction on mitigation, thereby deepening the adaptation burden on developing countries.

While President Xi alludes only to insufficient CC effort, Zuma explicitly mentions inaction as well as its consequences for developing countries, thereby reinforcing the moral weight of his country's negotiating position. The scarcity of references to the issue of continued emissions increases, despite its crucial importance to the fight against global climate change, can serve to illustrate how tacit knowledge plays a central part in the production and reception of texts. In other words, the absence of emissions increases does not indicate the issue's irrelevance but rather indicates its high contextual salience, as it constitutes the very reason for the conference in Paris

Reaction 2: Unilateral Measures and COP21

This stage of the narrative constitutes the response to the second complication—i.e., the problem of continued increases in CO_2 emissions, in spite of international diplomatic efforts. In his statement, President Obama praises the individual efforts by UNFCCC members, setting a collaborative tone:

> (15) USA: Already, prior to Paris, more than 180 countries representing nearly 95 percent of global emissions have put forward their own climate targets. That is progress.

It is worth noting that all four leaders mention their respective states' unilateral measures, thereby constructing their countries as heroes and reinforcing their arguments and negotiating positions. The US statement highlights investments in wind and solar power and reductions in emissions over the previous seven years, while China talks of its considerable share in renewable energy capacity and utilization, as well as its leading role in energy conservation, also emphasizing its current and future efforts to aid developing countries in confronting climate change. India focuses on levies and subsidy reductions to transition away from fossil fuels, and South Africa, while making no specific mention of current efforts, states that the Group of 77 and China "take our responsibilities seriously and are already making meaningful contributions within our respective capabilities."

Furthermore, several of the statements offer up concrete goals for future emissions reductions, except for South Africa, which instead refers to previously published plans, characterized as an "ambitious" Intended Nationally Determined Contribution (INDC).[3]

> (16) China: In its Intended Nationally Determined Contributions, China pledges to peak CO_2 emissions by around 2030 and strive to achieve it as soon as possible, and by 2030, reduce CO_2 per unit of GDP by 60–65% over the 2005 level, raise the share of non-fossil fuels in primary energy consumption to about 20% and increase forest stock by around 4.5 billion cubic meters over 2005. This requires strenuous efforts, but we have confidence and resolve to fulfill our commitments.

> (17) India: So, we have set ambitious targets. By 2030, we will reduce emissions by 33–35% of 2005 levels, and produce 40 per cent of our power from non-fossil fuels. We will achieve it by expanding renewable energy—for example, by adding 175 Gigawatts of renewable generation by 2022. We will enlarge our forest cover to absorb at least 2.5 billion tonnes worth of CO_2.

> (18) USA: For our part, America is on track to reach the emissions targets that I set six years ago in Copenhagen—we will reduce our carbon emissions in the range of 17 percent below 2005 levels by 2020. And that's why, last year, I set a new target: America will reduce our emissions 26 to 28 percent below 2005 levels within 10 years from now.

The enumeration of current efforts and promises for the future contributes to the formation of the hero role, particularly in the case of China's detailed account of its efforts to aid developing countries. Such descriptions lend moral support to what constitutes the bulk of the statements and the heart of the narratives—namely, the hopes and appeals for a comprehensive agreement at the negotiations in Paris. This part is a heterogeneous set of proposals, some of which can be found in several of the four statements. In the following, I will present an overview of these proposals, before looking more closely at the linguistic and discursive properties of those that set the statements apart and that reflect a strategic positioning on the part of the speaker.

Table 3.1 Propositions by the Four Leaders

Propositions	*Countries*			
Common but differentiated responsibilities	China	India	South Africa	USA
Adaptation as part of climate deal		India	South Africa	
Agreement on loss and damage		India		
Capacity building for developing countries	China	India	South Africa	USA
The need to include private sector, including business	China			USA
Doing away with zero sum game mentality	China			
No end to conventional energy		India		
COP21 as the deciding moment in the fight against CC		India	South Africa	USA

The primary points of negotiation—namely, common but differentiated responsibilities and capacity building for developing countries—are present in all four statements. Furthermore, even though CC adaptation is not explicitly advocated as a part of an agreement by China or the United States, their commitment to the issue is highlighted in their presentations of ongoing international partnerships. Moving on to the points that set each statement apart, the particularity of the Chinese text is that it rejects the notion of a "zero sum game," thereby alluding to previous failures to reach an agreement:

> (19) China: We should create a future of win-win cooperation, with each country making contribution to the best of its ability. For global issues like climate change, a take-more-give-less approach based on expediency is in nobody's interest. The Paris Conference should reject the narrow-minded mentality of "**zero sum game**" and call on all countries, the developed countries in particular, to assume more shared responsibilities for win-win outcomes.

By talking of a "narrow-minded mentality of 'zero sum game,'" President Xi rejects not only a point of view (i.e., 'negotiations on CC are a zero sum game') but also a narrative constructed around this point of view. In this rejected narrative, the complication is not CC but rather the imposition of negotiations that threaten economic interests. The reaction is then to work for narrow self-interests, while the resolution is to minimize the economic costs of any agreement. This is the closest that any of the statements have come to admonishing developed countries for lack of good will in previous efforts to curb GHG emissions, thereby portraying them as deceitful villains beyond the scope of historical responsibilities (see above).

In his remarks, President Obama made a similar point:

> (20) One of the enemies that we'll be fighting at this conference is cynicism, the notion **we can't do anything about climate change.**

The point of view that is attacked here is 'we cannot do anything about climate change', which represents a kind of cynicism that is about not self-interest but fatalism. Contrary to President Xi's remarks, the rejection of such a fatalist narrative does not necessarily imply a deceitful villain.

Moving on to Prime Minister Modi's statement, two elements stood out as unique—namely, loss and damages (21) and the need for conventional energy (22):

> (21) India: Equity means that national commitments must be consistent with the carbon space nations occupy. We also need a strong agreement on adaptation and loss and damage.

Loss and damage is defined by the UNFCCC as "the actual and/or potential manifestation of impacts associated with climate change in developing countries that negatively affect human and natural systems" (UNFCCC 2012). Given that it was a point of contention between developed and developing countries ahead of COP21, the absence of any argument supporting the assertion in example (21) is conspicuous.

> (22) India: We still need conventional energy. We should make it clean, **not** impose an end to its use.

Through the use of negation, Modi refutes an implicit point of view: 'We should impose an end to the use of conventional energy'. The voice behind this point of view could correspond to a collective representing most of the UNFCCC members, meaning that India positions itself in opposition to its negotiating partners on a significant point. The backdrop of this position is the fact that 300 million Indians still lack electricity and that the government has announced that India will double its coal output by 2020, with the goal of providing electricity to all by 2022 (ibitimes.co.uk, Dec. 1, 2015).

South Africa's statement contains no unique propositions regarding COP21, but its argumentation merits a closer look. As the leader of the

Group of 77 and China, South Africa emphasizes the moral imperative of prioritizing poverty eradication above emissions reductions, then adding the following point:

> (23) South Africa: We should also realise that developing countries now have the greatest mitigation potential, particularly in larger developing countries. Not providing finance, technology and capacity-building support to these efforts will inevitably lead to us falling far short of our goal to limit temperature increase to less than two degrees Celsius. If Paris is to succeed, the concerns of the Group must **therefore** be heard.

In this excerpt, President Zuma argues for finance, technology, and capacity building as a practical imperative, adding to the moral dimension of this issue. In this way he constructs capacity building as a necessary condition for the transition from secondary complication to resolution, thereby placing the responsibility for mitigation in developing countries on developed countries. The logic of this reasoning rests on the connective *therefore*, which constructs a major premise linking the minor premise to the conclusion:

> **Minor premise**: The two-degree target depends on capacity building.
> **Major premise** (collective voice): Meeting the two-degree target is a criterion for success.
> **Conclusion**: Concerns regarding capacity building must be heard.[4]

As in example (6), the force of the implicit major premise is that it is constructed by the recipients themselves during interpretation, which contributes to its argumentative force.

The distinguishing feature of Obama's portrayal of the Paris negotiations is his strong emphasis on its significance:

> (24) USA: What should give **us** hope that this is a turning point, that this is the moment **we** finally determined we would save our planet, is the fact that **our nations** share a sense of urgency about this challenge and a growing realization that it is within our power to do something about it.

In example (24), Obama constructs collective voices around the pronouns *us* and *we*, which includes leaders and negotiators, and *our nations*, which lends legitimacy to the effort. These voices have several points of view:

> *We* (leaders and negotiators at COP21): 'This is a turning point', 'This is the moment we will save our planet'.
> *Our nations* (a global collective voice): 'It is urgent', 'It is within our power to do something about it'.

The particularity of Obama's portrayal of COP21 is not so much in the strategic choice of measures to advocate but rather in its dramatization. By underlining the significance of the event, the statement strengthens the secondary reaction and its pivotal role in the narrative arc. In other words, the stakes of the moment are underlined and reinforced through Obama's statement. Furthermore, this secondary reaction is constructed as belonging to collective voices in example (24), thereby increasing its credibility. As we will see, the stakes are further underlined through the statement's detailed description of the long-term consequences in the case of failure in Paris (see example (29) below).

The historical significance of COP21 is also mentioned in the Indian and South African texts, but to a lesser extent:

> (25) India: Over the next few days, we will decide the fate of this planet. We do so when the consequences of the industrial age powered by fossil fuel are evident, especially on the lives of the poor.

> (26) South Africa: This Paris Conference is a crucial juncture in our fight against climate change. We owe it to ourselves and future generations to rise to the occasion and conclude a meaningful agreement here that will enhance collective, ambitious and equitable international action.

Resolution and Final Situation: A New Direction and a Sustainable Future

The resolution of the narrative constitutes the fulfillment of the propositions outlined in the secondary reaction, in the months and years following COP21. Previous studies have indicated that government policy texts on CC tend not to elaborate on resolutions and final situations, which are only presented briefly as potential future outcomes of policy proposals (see Fløttum & Espeland 2014; Fløttum & Gjerstad 2013a, 2013b). This is also the case in the Chinese, Indian, and South African statements. Example (27), taken from the closing remarks of the Chinese statement, is representative of this:

> (27) China: Let us join hands to contribute to the establishment of an equitable and effective global mechanism on climate change, work for global sustainable development at a higher level and bring about new international relations featuring win-win cooperation.

This example is part of the secondary reaction, while the resolution is restricted to noun phrases such as "an equitable and effective global mechanism on climate change." To properly account for the resolution component and its relation to the rest of the narrative in example (27), the analysis would need to consider the narrative composition of the sentence, which requires a more detailed linguistic perspective (for examples of such analyses, see

Fløttum and Gjerstad 2013a; Fløttum and Gjerstad 2016). However, the essential point here is the fact that the latter stages of the narrative are only alluded to, a succinctness that contrasts with the elaborate nature of the secondary reaction. This is in contrast to the US statement, which on the contrary explains how a successful resolution will offer few victories, before painting a picture of the final situation, in which future generations are cast as characters looking back at the heroic accomplishments of those who took part in the negotiations in Paris:

> (28) USA: And, my fellow leaders, accepting this challenge will not reward us with moments of victory that are clear or quick. Our progress will be measured differently—in the suffering that is averted, and a planet that's preserved. And that's what's always made this so hard. Our generation may not even live to see the full realization of what we do here. **But** the knowledge that the next generation will be better off for what we do here—can we imagine a more worthy reward than that? Passing that on to our children and our grandchildren, so that when they look back and they see what we did here in Paris, they can take pride in our achievement.

The two last stages of the narrative are tied together by the concessive connective *but*. According to the ScaPoLine, this marker signals that the speaker admits to the truth of a point of view before offering another POV that is presented as more relevant for the issue at hand (Nølke, Fløttum, and Norén 2004). In the present case, Obama admits to the POV that a successful resolution can only be understood in negative terms—i.e., through the avoidance of disaster. Who is responsible for this point of view is a matter of contextual inference—in this case, it may be attributed to all those who are involved in the negotiations. Furthermore, the points of view that are tied together by the connective serve as arguments for two opposing implicit conclusions, the interpretation of which is also a matter of contextual inference. The following is one possibility:

Argument 1 (unknown voice): "Accepting this challenge will not reward us with moments of victory that are clear or quick. Our progress will be measured differently—in the suffering that is averted, and a planet that's preserved. And that's what's always made this so hard. Our generation may not even live to see the full realization of what we do here." → "Success won't be appreciated."

Argument 2 (speaker): "The knowledge that the next generation will be better off for what we do here—can we imagine a more worthy reward than that? Passing that on to our children and our grandchildren, so that when they look back and they see what we did here in Paris, they can take pride in our achievement." → "Success will be appreciated."

Alternate Resolution and Final Situation: Business as Usual Leading to Ecological and Human Disaster

In addition to being alone in elaborating on the resolution and final situation, Obama's statement is also the only one that offers a description of the major disruptions that will follow if COP21 fails, inspired by the changes that the President observed in Alaska during the summer (see example (9) above):

> (29) USA: And it was a preview of one possible future—a glimpse of our children's fate if the climate keeps changing faster than our efforts to address it. Submerged countries. Abandoned cities. Fields that no longer grow. Political disruptions that trigger new conflict, and even more floods of desperate peoples seeking the sanctuary of nations not their own. That future is not one of strong economies, nor is it one where fragile states can find their footing. That future is one that we have the power to change. Right here. Right now. But only if we rise to this moment.

By developing a vivid description of a disastrous hypothetical future, the statement enters into the realm of storytelling in the true sense of the word. The description is then used argumentatively to advocate strong action in Paris, giving way to the secondary reaction of the present—namely, COP21—and constructing it as the deciding moment for humanity. In this way, the leaders and negotiators are attributed a deciding role in the story on human survival in the face of climate change.

Concluding Remarks

Using narrative analysis helps to identify the underlying story of policy texts and helps to identify how different storylines resemble and differ from one another. The aim of the present chapter was to highlight strategically significant aspects of leader statements at the start of COP21 in Paris, using a combined narrative and polyphonic analysis. How do the delegations of the world's two largest emitters and those of newly industrialized nations represent the problem of climate change and its potential solutions in their statements? In the narrative about climate change, who are presented as the main actors—i.e., the heroes, villains, and victims—and, more specifically, what are the roles of various countries according to the different statements? Do the narratives on CC reflect underlying national and international inequities as well as conflicts regarding historical responsibilities?

The narrative structures of the four statements are above all characterized by their similarities, both in their approach to historical responsibilities and in their shared emphasis on the secondary reaction (unilateral measures and COP21). While there are differences regarding the specific proposals for the negotiations during the Paris Conference, there is agreement on what is widely regarded as the two most essential issues—namely, common but differentiated responsibilities and capacity building. To the extent that there are any considerable differences regarding COP21, India stands out in advocating an agreement on loss and damage and in its refusal to agree to an end to conventional energy. In their cast of characters, the narratives are also highly similar, each country casting themselves as heroes by presenting national initiatives, and the UNFCCC as potential heroes, provided they can successfully negotiate a deal. Interestingly, victims do not feature prominently in the stories. Obama talks of the existential threat that CC poses to island nations, while the world's poor appear as victims only twice, in Zuma's and Modi's statements. When comparing the four statements, the main difference is not in policy content but rather in their style, and the US statement stands out in more clearly resembling a story than the other three. It dramatizes COP21 as a pivotal point in the climate change narrative, and it paints vivid pictures of possible worlds in the case of both success and failure. As such, the statement appears to be more motivational than policy oriented. In a broader sense, the two alternate future perspectives of Obama's remarks show the argumentative force of political storytelling. By drawing causal links between present actions and future impacts in emotional and value-laden terms, storytelling becomes argument, indicating the persuasive potential of policy narratives.

Notes

1 For a presentation of enunciative linguistics, see Sophie Marnette, 2005: *Speech and Thought Presentation in French: Concept and Strategies*. John Benjamins. Amsterdam—New York.

2 Previous research has indicated that policy narratives may be condensed into single sentences, meaning that lexis and syntax are relevant units of analysis when considering the micro-level (Fløttum and Gjerstad 2013a, 2013b).

3 INDCs are the post-2020 climate actions that the parties had publicly declared ahead of the negotiations in Paris (http://www.wri.org/indc-definition). See also chapter 8 in this volume.

4 The underlying logical relation between the arguments is transitive and can be resumed as follows: 'the two-degree target depends on capacity building'—'success depends on the two-degree target'—'success depends on capacity building'.

References

Adam, J.-M. 1992. *Les textes: Types et prototypes*. Paris: Nathan.

Adam, J.-M. 2008. *La linguistique textuelle. Introduction à l'analyse textuelle des discours*. Paris: Armand Colin.

Fløttum, K. 2010. "A linguistic and discursive view on climate change discourse." *Asp-Revue du GERAS* 58: 19–37.

Fløttum, K. 2013. "Narratives in reports about climate change." In Gotti, M., and C. S. Guinda (eds.). *Narratives in academic and professional Genres*, pp. 277–292. Bern: Peter Lang.

Fløttum, K., and T. J. Espeland. 2014. "Norske klimanarrativer—hvor mange 'fortellinger'? En lingvistisk og diskursiv analyse av to norske stortingsmeldinger." *SAKPROSA* 6 (4): 1–18.

Fløttum, K., and Ø. Gjerstad. 2013a. "Arguing for climate policy through the linguistic construction of Narratives and voices: The case of the South-African green paper National Climate Change Response'." *Climatic Change* 118 (2): 417–430.

Fløttum, K., and Ø. Gjerstad. 2013b. "The role of social justice and poverty in South Africa's national climate change response white paper." *South African Journal on Human Rights* 29 (1): 61–90.

Fløttum, K., and Ø. Gjerstad. 2016. "Narratives in climate change discourse." *Wiley Interdisciplinary Reviews: Climate Change*. 2017, 8:e429. doi: 10.1002/wcc.429

Gjerstad, Ø. 2013. "Voices and identities: The polyphony of Tony Blair and Nicolas Sarkozy at the European Parliament." In: Fløttum, K. (ed.). *Speaking of Europe: Approaches to complexity in European political discourse*, pp. 111–134. Amsterdam: John Benjamins Publishing Company.

International Business Times. 2015. COP21: Narendra Modi says solar power is key to reversing climate change. December 1. http://www.ibtimes.co.uk/cop21-narendra-modi-says-solar-power-key-reversing-climate-change-1531233

Jones, M. D. 2014. "Cultural characters and climate change: How heroes shape our perceptions of climate science." *Social Science Quarterly* 95 (1): 1–39.

Jones, M. D., M. K. McBeth, and E. A. Shanahan. 2014. "Introducing the narrative policy framework." In: Jones M. D., E. A. Shanahan, and M. K. McBeth (eds.). *The science of stories: Applications of the narrative policy framework in public policy analysis*, pp. 1–25. Basingstoke & New York: Palgrave MacMillan.

Marnette, S. 2005. *Speech and thought presentation in French: Concept and strategies*. Amsterdam & New York: John Benjamins.

Nølke, H., K. Fløttum, and C. Norén. 2004. *ScaPoLine: La théorie scandinave de la polyphonie linguistique*. Paris: Kimé.

UNFCCC. 2012. A literature review on the topics in the context of thematic area 2 of the work programme on loss and damage: a range of approaches to address loss and damage associated with the adverse effects of climate change. (FCCC/SBI/2012/INF.14) http://unfccc.int/resource/docs/2012/sbi/eng/inf14.pdf

UNFCCC. 2015. Leaders event at the Paris climate change conference 2015—COP 21/CMP 11. http://unfccc.int/meetings/paris_nov_2015/items/9331.php

4 Stories about Climate Change

The Influence of Language on Norwegian Public Opinion

Michael D. Jones, Kjersti Fløttum, and Øyvind Gjerstad

Story: A Linguistic and Social Scientific Approach

Many philosophers, theologians, playwrights, novelists, and academics have pondered the importance of stories and storytelling. For example, famed psychologist Jerome Bruner contends that "we seem . . . to have some predisposition, some core knowledge about narrative from the start" (Bruner 2003, 33). That is, human beings seem almost born with an intuitive sense of story, and stemming from this intuition is a compulsion to both understand and explain the world in story form. These individual storied expositions then congeal in a sociological sense (Polkinghorne 1988, 14), imbuing groups and collectives with shared understandings. Of course these sociological group-based stories become shared understandings of the objects and relationships that comprise the world as we know it (Campbell and Moyers 2011, 4). The stories thus transmit culture and understanding from person to person, underwriting reality while also modifying it as generations of human beings make their way through time. While nearly all—from lay people to presumed experts on narrative—would agree on the importance of story, unpacking how stories work, why, or even what they are has proven a herculean task producing incredibly diverse answers.

Narrative has been studied in-depth within many academic disciplines. Marketing research shows that narrative selling techniques are more likely to sell products than price-point advertising techniques (Mattilla 2000, 2002). Findings in psychology demonstrate that the more that individuals are transported into a story (suspend their interaction with the world around them and "fall" into the story), the more persuasive the story (Green and Brock 2005). Research in neuroscience has identified the centers for the brain responsible for narration (Troiani *et al.* 2006) and found that damage to those areas, such as that caused by Alzheimer's or advanced alcoholism, can be more severe than damage to other areas of the brain as individuals actually lose their sense of self when they lose the ability to narrate (Young and Saver 2001). Findings in political science reveal that narrative structures actually influence how individuals cognitively organize newly acquired information (Berinsky and Kinder 2006; Jones and Song 2014). No doubt theories about what narratives are and how they work are ubiquitous, as are findings. This chapter leverages two narrative approaches to explore the effect of cultural narratives on individual Norwegian preferences and perceptions of risk as they relate to climate change: the linguistic approach to narrative and the Narrative Policy Framework (NPF).

The linguistic approach to narrative is inspired by a long tradition beginning with the work of the Russian folklorist and scholar Vladimir Propp, who in the 1920s analyzed the basic plot components of Russian folk tales (translated to English in 1958, *Morphology of the Folktale*). From this literary and cultural position, the narrative perspective entered forcefully into analyses of nonfictional texts with the breakthroughs of text linguistics (e.g., Adam 1992; Van Dijk 1980; Werlich 1976). Our main inspiration in the present chapter is the classical five-part structure of a narrative as organized through five main components—*initial situation*, *complication*, *reactions*, *resolution*, *final situation*—which may be accompanied by a moral judgment (Adam 1992, 2008). This prototypical narrative sequence, through its five components, can be illustrated as follows (Adam 2008, 145):

> (*Initial situation*) Three young friends went swimming in the Annecy lake. (*Complication*) One of them, Janinetti, went missing. (*Reaction*) The others dove in. (*Resolution*) They managed to pull her out, (*Final situation*) but she was already dead.

At the heart of the narrative is the intrigue, or plot, a tension between the different actions, events, and actors that drive the account. In the above example, the idyllic trip turns into a crisis, prompting the characters' attempt to counteract the disappearance of their friend, but to no avail.

This text linguistic perspective has been developed for analyzing various types of texts related to climate change (Fløttum 2013; Fløttum and Gjerstad 2013a, 2013b, 2016) and is in the present study combined with the NPF, which takes the view that a policy narrative has *a setting*, *characters* (heroes, villains, and victims), *a plot*, and *a moral of the story* (Jones, McBeth, and Shanahan 2014, 5–7). This combination requires a conceptual clarification, as the compatibility between the two approaches is not immediately apparent. The NPF *setting* is the context in which a narrative takes place, corresponding to facts that are taken for granted, such as legal constraints, nation-state boundaries, demographics, and the like. The *plot* constitutes the temporal succession of events, from beginning through middle to end, tying characters to each other and to the setting, while the *moral* corresponds to the policy solution—e.g., nuclear power as a solution to climate change. Together these three notions correspond to the five-part structure of the prototypical narrative sequence, the difference being not in their overall content but in the way they relate to each other internally. While the NPF locates the temporality of the story within the notion of *plot*, temporality constitutes the main distinguishing feature of Adam's five components: the initial situation is followed by the complication, the reaction, the resolution, and the final situation. Conversely, these five components each comprise the setting as a contextual backdrop, though perhaps most prominently in the initial situation and complication. As for the moral, it is present in the reaction, resolution, and final situation.[1] While the two approaches are not perfectly congruent, they do offer considerable overlap, and, we argue, the differences are theoretically complementary. Therein is the benefit of employing both approaches simultaneously—in the construction of the narratives, the NPF has provided the macro-view of the plot and characters based on the cultural types of Cultural Theory (see below), while the text linguistic perspective has let us calibrate the different narrative components by adjusting the relation between sentences and constituents within sentences. We discuss this further in our analyses later in the chapter. Our next section details the approach to culture that we leverage for this study and how it might be applied to understand climate change in a Norwegian context.

Cultural Climate Change Stories

Cultural Theory (CT) is an approach to understanding culture initially developed by the late Mary Douglas and Aaron Wildavsky (1982). CT attempts to explain "how individuals confer meaning upon situations, events, objects, relationships—in short, their lives," or their cultures (Thomspson, Ellis, and Wildavsky 1990, xiii). CT posits two salient dimensions for determining individual and group cultures: grid and group. Group denotes affinities for socializing and participating in groups, while grid denotes the extent to which those groups prescribe and bind behaviors, beliefs, and attitudes. Intersecting the two dimensions produces a two-by-two typology, or four cultural types where each signifies a specific worldview. The *individualist* has both low group and low grid, thus shunning groups and preferring to operate with minimal constraints. The world they prefer to live in consists of unfettered markets and open competition. Adhering to this worldview, you might find the crafty business entrepreneur. In contrast, the *hierarch* has both high group and high grid. These individuals operate in tightly prescribed groups, where boundaries and expertise are valued—people are expected to play their roles and play them well. Within the hierarch worldview you might find the disciplined soldier or the highly orthodox bishop. *Egalitarians* operate with low grid but high group. These individuals prefer groups, but a particular kind of group where relationships within the group are flat—or equal. Divisions within the group play a minimal role. Within this worldview you might find a member of a commune. Finally, there is the *fatalist*. Having low group and high grid, these individuals find group membership to be difficult if not impossible to obtain. Due to their high grid, their lives are tightly prescribed and directed by others. For the fatalist, fate is capricious and their lives often prisons.

CT has been applied to study individuals, groups, and organizations across an array of academic disciplines and substantive topic areas (see Swedlow 2011). It has also been used extensively to study climate change, particularly the relationship cultural narratives have with different cultural types. A series of studies in the 1990s identified three main cultural narratives about climate change commonly associated with the three most commonly studied cultural types: egalitarian, individualist, and hierarch. Communicated by both individuals and institutions (e.g., media), these generic climate change stories are argued to be pervasive throughout communities (see Rayner and Malone 1998; Verweij *et al.* 2006). The egalitarian story, seeing nature as fragile, paints overconsumption as the cause of climate change; the villains are corporations and governments driving the consumption; and heroes of this story are environmental groups who combat overconsumption. The hierarch narrative paints mismanaged systems as the cause of climate change. For hierarchs, the villain tends to be incompetence and those who represent it, while the hero demonstrates organization and expertise such as government agencies, when portrayed in a positive light. Finally, we have the individualists. This group is the most likely to fall into the climate change denier camp. For the individualist, the cause of climate change is almost irrelevant, as whatever needs to be done about it will be solved by free markets. Villains for this group are those that impede free markets, while heroes advocate on the markets' behalf.

Recent research leveraging experimental survey designs has explored how the cultural narratives mentioned above interact with different cultural world views (Jones 2011, 2014a, 2014b; Jones and Song 2014). All of these studies rely on the same survey data and experimental research design, exposing a US representative sample of subjects to one of the three cultural narratives detailed above or a control list of facts about climate change. Findings from this research are several. Jones (2014b) observes that the better the cultural narratives were able to help respondents picture the events in the narrative, the more persuasive the narrative. Jones and Song (2014) analyzed data acquired after respondents were exposed to the narratives or the control that asked them to sort 27 different phrases or words from the cultural narratives into categories of their choosing—i.e., to group the words or phrases in ways that the respondents thought were meaningful. What they found was that when respondents were exposed to culturally congruent narratives (i.e., an egalitarian views the egalitarian narrative or an individualist views and individualist narrative), they tended to organize their concepts in the same way as the story (e.g., heroes, villains, etc.). However, if respondents were exposed to an incongruent narrative or the control treatment list of facts (e.g., an individualist that views an egalitarian narrative), then they organized words and phrases conceptually—all adjectives together, all groups together, etc. The short of it is that when narratives come from one's tribe, they tend to be remembered as the story was told. Finally, Jones (2014a) looked at responses from these same experiments that focused on the role of the characters in persuading. While observing no consistent role for villains, Jones did find that heroes seemingly drive cultural narrative persuasion. Respondents across all treatments were more likely to respond positively to the hero, regardless of their cultural type; moreover, the more they liked the hero character, the more likely they were to buy into the assumptions, arguments, and policy solutions offered in the cultural narratives. The research presented here further explores the hero effect observed in Jones (2014a) by exploiting two prominent approaches to the study of narrative: the NPF (Jones and McBeth 2010) and the methodological text linguistic framework (Fløttum 2013; Fløttum & Gjerstad 2013a, 2016) in conjunction with CT.

Research Design, Data, and Method

To assess the influence of cultural narratives on climate change opinion, this research leverages an experimental design built into a representative survey of Norwegian citizens to examine the following research question:

> **RQ:** To what extent do the characters (heroes and villains) within climate change narratives have a persuasive effect on perceptions of risk and willingness to take action?

The online quasi-experimental design consists of two experimental tracks built into the Spring 2015 Norwegian Citizen Panel Survey, wave 4. The sample is a representative sample of the Norwegian population where sample parameters are fitted to existing demographic characteristics of the Norwegian population.[2] While representative sampling is not the gold standard of a truly randomized sample, previous research has found representative samples comparable to random (e.g., Berrens *et al.* 2003; Best *et al.* 2001), and for purposes where the research team is primarily interested in theoretical relationships between variables and not necessarily characteristics of the sampled population (as we are here), then representative samples are appropriate. The data were collected from March 8 to April 9, 2015; 335 respondents completed the survey. We discuss the sample in more detail below.

Experimental Design

All respondents that participated in the experiment were first exposed to several pretreatment survey questions. These questions included several control items assessing the respondent's age, gender, education, and ideology. Among these questions were also a battery of six questions assessing the respondent's cultural proclivity toward egalitarianism and individualism (three individualistic measures and three egalitarian measures),[3] the two types of worldviews operationalized from CT within our experiments. Table 4.1 lists the descriptive statistics for each of the pretreatment variables used in our analysis.

Table 4.1 Pretreatment Variables, Sample Descriptive Statistics

Variable	*Measurement*	*N*	*Mean*	*Std. Dev.*
Individualism (Composite)	***Sum I1–I3*** *(Cronbach's Alpha .600)* 3 (Weak Individualist)—21 (Strong Individualist)	320	12.38	3.37
I1: Everyone should have equal opportunities for succeeding or failing without the authorities becoming involved.	1 (Strongly Disagree)—7 (Strongly Agree)	322	4.43	1.46
I2: Market competition is almost always the best way of meeting people's requirements.	1 (Strongly Disagree)—7 (Strongly Agree)	324	3.71	1.53
I3: If the system is fair, the cleverest people should earn the most money.	1 (Strongly Disagree)—7 (Strongly Agree)	327	4.24	1.53
Egalitarian (Composite)	***Sum E1–E3*** *(Cronbach's Alpha .678)* 3 (Weak Egalitarian)—21 (Strong Egalitarian)	325	14.56	3.66
E1: The world would be more peaceful if wealth was more evenly distributed between countries.	1 (Strongly Disagree)—7 (Strongly Agree)	328	4.86	1.64
E2: Society needs more justice so that wealth can be distributed more evenly.	1 (Strongly Disagree)—7 (Strongly Agree)	325	4.78	1.47
E3: I support tax reforms which would make companies and people on high incomes pay more tax.	1 (Strongly Disagree)—7 (Strongly Agree)	326	4.74	1.59
Ideology	0 (Far Left)—10 (Far Right)	302	4.76	2.23
Gender	0 (Female) or 1 (Male)	331	.5257	.500
Education	1 (No Completed Ed.) to 14 (Ph.D.)	321	8.25	3.31
Age	1 (18–25) to 7 (75 and above)	331	3.97	1.63

The Cultural Narrative Treatments

After completing the pretreatment survey questions, respondents were then randomly assigned to one of two experimental treatments designed specifically for the Norwegian context (see the Appendix for the English translations of the treatments used for this study). As such, the two experimental treatments were designed to account for different cultural characterizations of climate change within Norway. Consequently, we populated the content of each of the treatments with specific cultural content that approximated the individualist and egalitarian narratives previously discussed in the chapter. These two cultural types were chosen for two reasons: (1) past research indicated that persuasive effects with these two types of cultural narratives were more pronounced than the hierarch narrative (e.g., Jones 2010; Jones and Song 2014), and (2) resource constraints limited the number of experimental tracks we were able to field. In terms of imbuing the individualist narrative with cultural content, the narrative treatment focused on such themes as business, markets, and "buying and selling." On the other hand, the egalitarian narratives emphasized community by using phrases such as "way of life" and equality by invoking phrases such as "just distribution of wealth" to emphasize core aspects of the egalitarian worldview. The experiments were fielded in Norwegian; narratives and questions presented to the respondents are translated to English in this chapter.

Finally, and in the interest of our experimental design, we needed to identify specific narrative elements, where some elements could be varied while others were held constant to help us assess variation of specific facets of the narrative. Or, more simply, we needed to be able to assess what parts of the narrative have what effect. To operationalize narrative structure within our experimental treatments, we rely on both the NPF (Jones, McBeth, and Shanahan 2014; McBeth, Jones, and Shanahan 2014) and the methodological text linguistic framework developed by Fløttum and Gjerstad (Fløttum 2013; Fløttum and Gjerstad 2013a, 2016). The NPF is the same framework used in the studies conducted by Jones and collaborators to assess the effects of climate change cultural narratives in the United States mentioned earlier in the chapter. The NPF operationalizes narrative by disaggregating narrative into four distinct elements: setting, characters, plot, and moral of the story (or policy solution) (McBeth, Jones, and Shanahan 2014). From the methodological text linguistic framework, we operationalize analogous narrative concepts including primary and secondary *complications*, primary and secondary *reactions*, and *moral resolution*. This disaggregation of narrative allows us to construct our treatments in a way that holds some elements constant, while allowing us to vary others and assess variance in the experimentally manipulated elements. It also allows for some comparability between the study here and the previous work conducted by Jones (2014a, 2014b) and Jones and Song (2014). Table 4.2 lists the narrative elements of the cultural narratives on climate change within each of the treatments.

Table 4.2 Narrative Elements within the Cultural Narrative Treatments

Freq.	*Description*	*Setting*	*Hero*	*Villain*	*Victim*	*Moral*
147	Individualistic-Negative	Changing Climate	Green World	Norwegian Business	Citizens and Business	Cap and Trade
184	Egalitarian-Negative	Changing Climate	Norwegian Business	Green World	Citizens and Business	CO_2 Tax

The setting, or complication, of each narrative portrays likely changes caused by greenhouse gas emissions (GHGs)—increases in rainfall and sea level rises—as extreme or dramatic threats to Norwegians. The setting is held constant across the two conditions. The narratives portray the victims of these changes as Norwegian citizens and businesses. The victims too are held constant across the narratives. To play the hero and villain roles in our egalitarian and individualist narratives, a cast of characters was selected that prior literature has suggested would be likely to provoke emotional responses in their respective cultural types. Jones (2014a) and Jones and Song (2014) suggest that environmental characters are likely to be well received by egalitarians, and businesses are likely to be well received by individualists; conversely, the literature also suggests the inverse is true: egalitarians are likely to assign negative affect to business characters, and individualists are likely to assign negative affect to environmental characters. The fictitious group *Green World* plays the hero in the egalitarian narrative and the villain in the individualist narrative; the fictitious character *Norwegian Business* plays the hero in the individualist narrative and the villain in the egalitarian narrative. To calibrate the narratives to their cultural worldviews, each also includes value-laden expressions, such as "radical" and "greedy," tailored for the values of the cultural type. Conversely, there are no linguistic markers of uncertainty (epistemic modality), in order to make causal links and responsibilities as clear as possible. Causality is explicitly marked through hypothetical structures (*if A, then B*), connectives (*thus, then*), and verbs (e.g., *exploit*, which implies intentionality). To ensure that the reader is able to follow the narrative, temporal succession follows the text, and both lexical and syntactic features are kept simple—the narratives contain only one subordinate clause each. Finally, the main tool to clearly mark the "twists and turns" of the narrative are counter-argumentative connectives. Through the use of *however* and *but*, the text turns first from the first reaction phase to the complication phase, and then to the second reaction phase. These two connectives thus serve as the pivot points of the plot, allowing for a very short narrative without compromising the clarity of the plot.

By way of example, we offer a discussion of the individualistic narrative to further illustrate how narrative theory was employed to construct our experimental treatments (also see the Appendix):

> ***(SETTING, PRIMARY COMPLICATION)*** *If the world continues with greenhouse gas emissions like today, changes in climate for Norway will include extreme increases in rainfall and dramatic sea level rises. This will constitute a major threat to people and their way of life across the country.* ***(PRIMARY REACTION)*** *The organization Norwegian Business* ***[HERO]*** *fights climate change by means of a quota trading system (Cap & Trade). This would let businesses reduce* CO_2 *emissions through buying and selling GHG emission allowances.* ***(SECONDARY COMPLICATION)*** *However, the radical environmental organization Green World* ***[VILLAIN]*** *pushes its own destructive agenda. It battles constantly to undermine the free market.* ***(SECONDARY REACTION)*** *But Norwegian Business* ***[HERO]*** *will expose undue political influence of environmental lobbying groups.* ***(MORAL, RESOLUTION)*** *It* ***[HERO]*** *emphasizes that Cap & Trade removes the need for more emission regulations that threaten growth. With the liberty to invest, businesses can keep the economy growing while reducing* CO_2 *emissions.*

The first sentence explicates the setting, in order to ensure that the text is accessible to respondents. The secondary complication introduces the villain and with it the political conflict. None of the narrative components exceed two sentences, in order to maintain simplicity and balance between them. The final situation is not developed in a sentence of its own but hinted at in the last sentence ("keep the economy growing"). This corresponds with analyses of authentic political texts on climate change (see Fløttum and Gjerstad 2013a, 2013b). Through the use of linguistics and the NPF, we were thus able to calibrate the narrative arc and ensure a high degree of precision in the composition of the experimental stimuli, which are very similar in composition and length (136 total words and 142 total words, respectively). The following section explores our primary research question via analysis of data obtained from the aforementioned experimental design.

Analyses and Findings

Past NPF studies have hypothesized that individuals will show positive affect for hero characters and negative affect for villain characters (e.g., Jones 2014a). Thus, to address our research question seeking to assess the role of characters in shaping opinion about climate change, we apply Ordinary Least Squares (OLS) regression analyses to the following hypotheses:

> H_{1A}: The policy narratives will facilitate positive affect for hero characters.
> H_{1B}: The policy narratives will facilitate negative affect for villain characters.

To assess the above hypotheses, we ran two discrete multivariate OLS regressions to determine if our primary independent variable, the narrative treatments, had any relationship with both hero and villain affect. For this test, we operationalized the narrative experimental treatments as binary dummy variables (1 for present, 0 for not), with the second omitted experimental narrative treatment removed from each of the two regression models, thus serving as a baseline within the equation (reflected in the constant). Both of our character affect dependent variables (villains and heroes) are based on single-question posttreatment survey items asking respondents to assess the characters on a scale from 0 to 10, where 0 reflects a highly negative view and 10 indicates a highly positive affective assessment of the character (see Table 4.3). For both models, independent control variables include ideology (left/right), gender, education, age, and cultural worldview (individualist or egalitarian) (see Table 4.1 for variable descriptives). Table 4.4 summarizes the results for the individualist narrative and character affect. Table 4.5 summarizes the results for the egalitarian narrative and character affect.

Regarding Tables 4.4 and 4.5, our primary interests are the individualist narrative and egalitarian narrative independent variables and their relationship with the hero and villain dependent variable characters within each narrative. Table 4.4 shows a significant and positive (.675, $p < .05$) relationship between the hero character (*Norwegian Business*) and the individualist narrative. We also find that the coefficient for the *Green World* villain character is negative, but not significant within the individualist narrative.

Table 4.3 Posttreatment Dependent Variables, Descriptive Statistics, by Narrative Treatment

Variable	*Measurement*	*Individualist Narrative*			*Egalitarian Narrative*		
		N	*Mean*	*Std. Dev.*	*N*	*Mean*	*Std. Dev.*
Green World Affect	0 (Highly Neg) to 10 (Highly Pos)	118	4.64	2.23	154	4.94	2.75
Norwegian Business Affect	0 (Highly Neg) to 10 (Highly Pos)	120	4.38	2.24	156	3.82	2.23
Text Accuracy	0 (No Confidence) to 10 (Full Confidence)	133	4.15	2.45	166	4.66	2.55
Personal Risk	0 (No Threat) to 10 (Extreme Threat)	139	5.09	2.59	175	4.84	2.66
Sociotropic Risk	0 (No Threat) to 10 (Extreme Threat)	141	5.86	2.12	174	5.69	2.28
Regulate GHG's	0 (Not Important) to 10 (Extreme Importance)	142	7.23	2.43	172	6.66	2.93

Table 4.4 The Individualist Cultural Narrative and Character Affect

	Green World	*Norwegian Business*
Constant	5.848*	2.867*
	(1.282)	(1.130)
Individualist Narrative[a]	–.440 (Villain)	**.675*** **(Hero)**
	(.312)	**(.276)**
Egalitarianism	.106*	–.058
	(.050)	(.044)
Individualism	–.027	.087
	(.052)	(.046)
Ideology	–.310*	.200*
	(.094)	(.084)
Male	–.279	–.281
	(.311)	(.277)
Education	–.028	–.058
	(.046)	(.041)
Age	–.049	.078
	(.096)	(.085)
Adjusted R^2	.149	.131
F-statistic	7.020	6.191
N	240	242

*p <.05, standard errors reported in parentheses.
[a]= one-tailed test; otherwise, tests are two-tailed.

Table 4.5 The Egalitarian Cultural Narrative and Character Affect

	Green World	*Norwegian Business*
Constant	5.408*	3.542*
	(1.253)	(1.106)
Egalitarian Narrative[a]	.440 (Hero)	**–.675*** **(Villain)**
	(.312)	**(.276)**
Egalitarianism	.106*	–.058
	(.050)	(.044)
Individualism	–.027	.087
	(.052)	(.046)
Ideology	–.310*	.200*
	(.094)	(.084)
Male	–.279	–.281
	(.311)	(.277)
Education	–.028	–.058
	(.046)	(.041)
Age	–.049	.078
	(.096)	(.085)
Adjusted R^2	.149	.131
F-statistic	7.020	6.191
N	240	242

*p <.05, standard errors reported in parentheses.
[a]= one-tailed test; otherwise, tests are two-tailed.

For Table 4.5, we see similar directional results for the egalitarian narrative. That is, there is positive affect for the hero character and negative affect for the villain. However, only affect for the villain *Norwegian Business* is significant (–.675, p.<.05). Thus, while we do have significance in expected directions, it is not consistent for both narratives. Despite the confirmatory elements of the results, a conservative interpretation of the data insists that we fail to accept H_{1A} and H_{1B}.

One possible reason for our failure to conclude that the policy narratives are independently driving affective variation might be found in two of the control variables. In both models, egalitarianism is significantly associated with positive affect for *Green World*. Similarly, across both equations, we see that ideology has an influence on character affect. For both equations, the more to the right (conservative) a respondent's political leanings, the more the respondent dislikes *Green World* and the more he or she likes *Norwegian Business*. Simply put, character affect might be more closely associated with the cultural and ideological dressings that the respondent is able to pin on the character due to the text, as opposed to the narrative structure itself.

Next, we examine the relationship in our experiment with character affect and several persuasive elements of our cultural narratives. More specifically, Jones (2014a) found that positive hero affect in cultural narratives about climate change applied to a sample from the United States was positively associated with a number of arguments and assumptions made within the cultural narratives, including perceptions of risk, whether or not respondents believed climate change to be real, and respondents' willingness to take action. Our study examines similar propositions. Thus, regarding hero affect, we hypothesize the following:

- H_2: As affect for the hero character increases, the respondent's belief in the accuracy of the text will also increase.
- H_3: As affect for the hero character increases, the respondent's sense of threat that climate change poses will also increase.
- H_4: As affect for the hero character increases, the respondent's willingness to take action to control or regulate greenhouse gasses will also increase.

Similar to our last test, to assess H_2–H_4, we ran several multivariate OLS regression analyses. Our dependent variables of interest included several survey items that subjects responded to after exposure to the narrative experimental treatments. These survey items asked respondents to assess how accurate they thought the text was (*text accuracy*), how much of a threat they thought climate change posed to them personally (*personal risk*), how much of a threat they thought climate change posed to Norway more generally (*sociotropic risk*), and how willing respondents were to regulate greenhouse gasses in Norway (*regulate GHGs*). Table 4.3 above summarizes descriptive statistics and measurement approaches for each of these dependent variables. For all models, independent variables include ideology (left/right), gender, education, age, and cultural worldview (individualist or egalitarian). Table 4.6 summarizes OLS analyses for subjects exposed to the individualist narrative; Table 4.7 summarizes OLS analyses for subjects exposed to the egalitarian narrative. Our independent variable of interest is the hero character within each cultural narrative experimental treatment: *Green World* for the egalitarian narrative and *Norwegian Business* for the individualistic narrative.

Table 4.6 presents OLS regression analysis intended to help us assess the relationship between the hero in the individualistic narrative—*Norwegian Business*—and several dependent variables designed to capture the thrust of the cultural narrative about climate change, which includes the idea that climate change is real (*text accuracy*), that it is a threat (both personally and socially), and that something ought to be done about it (*regulate GHGs*). For two of these dependent variables, we see significant relationships between

Table 4.6 The Individualist Narrative: Character Affect, Text Veracity, Risk, and Willingness to Regulate GHGs

	Text Accuracy	*Personal Risk*	*Sociotropic Risk*	*Regulate GHG's*
Constant	.850	1.161	1.927	3.202
	(1.947)	(2.157)	(1.543)	(1.852)
Norwegian	**.695***	**.210**[b]	**.162***	.063
Business Hero[a]	**(.124)**	**(.134)**	**(.095)**	(.115)
Green World	.098	.224	.081	.187
Villain	(.110)	(.121)	(.087)	(.105)
Egalitarianism	.021	.151*	.184*	.145*
	(.069)	(.076)	(.055)	(.065)
Individualism	.070	–.131	–.118A	–.047
	(.084)	(.087)	(.062)	(.074)
Ideology	–.335*	.096	–.005	.004
	(.139)	(.145)	(.104)	(.125)
Male	.286*	.285	–.232	–.691
	(.459)	(.498)	(.353)	(.422)
Education	.001	.115	.096	.156*
	(.069)	(.075)	(.054)	(.064)
Age	.021	–.061	.238*	.123
	(.141)	(.150)	(.107)	(.128)
Adjusted R^2	.253	.073	.237	.186
F-statistic	5.096	1.976	4.915	3.919
N	97	99	101	102

*p <.05, standard errors reported in parentheses.
[a]= one-tailed test; otherwise, tests are two-tailed.
[b]= *p <.10.

hero affect and the dependent variables, in the expected positive direction. We see this positive and significant relationship with the respondent assessment that the text is accurate (*text accuracy*) and that climate change poses a threat to Norway (*sociotropic risk*). We do not, however, observe significance with either *personal risk* or willingness to *regulate GHGs*. Yet it is worth mentioning that while *personal risk* does not reach the .05 threshold of significance, it is quite close and in the hypothesized positive direction ($p < .10$).

Table 4.7 shows results from OLS regression analysis intended to help us assess the relationship between respondent affect for the *Green World* hero in the egalitarian narrative and respondent assessment of the accuracy of the text (*text accuracy*), the societal (*sociotropic risk*) and personal threat (*personal risk*) that climate change poses, and how willing they are to *regulate GHGs*. Each of these variables shows a positive and significant relationship with hero affect—as respondents' affect for *Green World* increases, so too does their assessment of the accuracy of the text, their sense of threat that

Table 4.7 The Egalitarian Narrative: Character Affect, Text Veracity, Risk, and Willingness to Regulate GHGs

	Text Accuracy	*Personal Risk*	*Sociotropic Risk*	*Regulate GHG's*
Constant	−1.704 (1.336)	5.034* (1.718)	3.648* (1.387)	7.518* (1.801)
Norwegian Business Villain	.181* (.073)	.123 (.091)	.203* (.074)	.061 (.095)
Green World Hero[a]	**.637*** **(.062)**	**.374*** **(.080)**	**.360*** **(.065)**	**.480*** **(.084)**
Egalitarianism	.065 (.053)	.039 (.068)	.122* (.055)	−.015 (.071)
Individualism	.145* (.052)	.062 (.068)	.036 (.055)	−.054 (.071)
Ideology	−.221* (.102)	−.245 (.130)	−.241* (.105)	−.354* (.137)
Male	.575 (.318)	−.727 (.409)	−.492 (.331)	−.453 (.430)
Education	.014 (.049)	−.025 (.062)	−.085 (.050)	−.023 (.065)
Age	.097 (.101)	−.468* (.132)	−.109 (.107)	−.092 (.138)
Adjusted R^2	.555	.314	.423	.371
F-statistic	21.074	8.601	13.110	10.648
N	129	133	132	131

*$p < .05$, standard errors reported in parentheses.
[a]= one-tailed test; otherwise, tests are two-tailed.

climate change poses (both personal and societal), and their willingness to regulate Norwegian GHGs.

Regarding H_2, where it was hypothesized that hero affect would be positively related to respondent belief that the text is an accurate depiction of climate change, we find that, for both the individualist and the egalitarian narratives, this hypothesis is confirmed. H_3 hypothesized that as hero affect increased, so too would the respondent's assessment of threat climate change poses. Here we have more mixed results. For both egalitarian and individualist narratives, this holds to be true for *sociotropic risk* but not *personal risk*. Our data show a positive and significant relationship between *personal risk* and hero affect in the egalitarian narrative but not in the individualist narrative. However, *personal risk* and hero affect in the individualist narrative does move in the expected positive direction and approaches significance. Taken in total, we find these data to be compelling enough to accept H_2, as three of the four relationships are significant and one is nearly so. Finally, regarding H_4, where we hypothesized that hero affect would have a positive relationship with regulating GHGs, we find our least compelling results. Only in the egalitarian narrative did we find evidence that hero affect was correlated with willingness to regulate GHGs—no relationship was observed in the individualist narrative. Given the split result, we do not find the evidence sufficient to support H_4.

Discussion and Conclusion

In total, our experiments have allowed us to reject several hypotheses, including H_{1A} and H_{1B}. Recall that the H_{1A} and H_{1B} hypotheses allowed us a closer look as to whether or not narrative structure and content play an important role in steering subject affective responses to characters within the narrative. While we found some evidence that the narrative structure and content of our experiments was directing hero and villain affect, it was not a uniform effect across experimental conditions. We conjectured that it was likely ideological and cultural attachments to the characters that were the conflating driver in our narratives, but we cannot be certain of our conjectures. In the short term, we must conclude that our data do not find the same narrative driver of character affect present in the Jones studies of the United States (e.g., 2010; 2014a). If our data are any indicator, then the generalized effect found in the Jones studies appears to be more contextual than originally thought, and other drivers of character affect should be accounted for in future studies.

Our next set of hypotheses examined to what extent positive affect for the hero character was responsible for positive responses to a number of dependent variables integral to the stories being told about climate change in our cultural narrative treatments. Key dependent variables for these tests include respondent assessments of the accuracy of the text, respondent assessment of personal and societal risk, and respondent willingness to regulate GHGs in Norway. Regarding all but the willingness to regulate GHGs, we found robust evidence that positive affect for the hero character in both the egalitarian and the individualist narratives coincided with an increased sense of risk about climate change (both personally and societally) and a higher assessment of the narrative's accuracy. However, we found that only in the egalitarian narrative did hero affect seem to influence respondents' willingness to act on climate change in terms of their willingness to regulate Norwegian GHGs. Thus, our findings are not quite as robust as Jones (2014a), where similar experiments were conducted with a US population sample. In that study, there were more dependent variables, and "in each and every case [. . .] more positive affect for the hero means higher respondent scores" across each (Jones 2014a, 22). Our findings, however, are quite confirmatory and do yield supporting evidence that hero affect is an important driver of narrative persuasion in climate change narratives—not just in the United States, but also in Norway.

Even though we cannot conclude that the policy narratives on climate change are independently driving affective variation, there is clear support for considering hero affect as an important driver of narrative persuasion. This finding was made possible by the research design, where the linguistic approach complements the NPF approach. One important element in this is the linguistic construction of the narratives with different forms of tension in which the hero characters were implicated. More concretely, this was done by calibrating the narratives in accordance with primary and secondary complications and reactions. Furthermore, the complications and reactions were reinforced by value-laden expressions tailored for the values of the respective cultural types and by the marking of tensions through counter-argumentative connectives, such as *but* and *however*. These two connectives, with a clear rhetorical effect, thus served as the pivot points of the plot.

We believe that the attention to syntactic and semantic details in experiments such as the one undertaken here contributes to a level of precision that is useful both when constructing narratives for experimental purposes and when analyzing the data. We envisage more fruitful collaboration between the NPF and the text linguistic approaches in research related to climate change discourse. From the linguistic side, the possibility of testing hypotheses regarding the interpretative and persuasive effects of linguistic markers seems particularly promising. In addition, the broader cultural perspective of NPF has the potential to link individual narratives to a larger political and societal context. Such complementary contributions can serve as the basis for more and broader cross-disciplinary research of climate change discourse, where linguistics and political science may bring in perspectives from both climate science and psychology.

Finally, the NPF has shown the relevance of the narrative perspective on issues other than climate change, which means that its collaboration with the text linguistic approach could be developed further. This expansion could be beneficial, as respondents' opinions on issues that have received little public attention might be less settled, meaning that narratives may have greater influence on respondents during experimental treatments. This would allow for further studies of the importance of narrative variables and linguistic features.

Notes

1 See Jones and McBeth (2010) for a more detailed discussion of the NPF's (and the academic discipline of Public Policy) relationship with narratology.

2 Panel members were recruited in wave 1 and wave 3. The samples in wave 1 and wave 3 were drawn from the "National Registry" of Norway. This register encompasses everyone born in Norway as well as former and current inhabitants. The Norwegian Tax Administration is responsible for this register but has partly outsourced the administration to the private IT company Evry. Evry drew the sample on behalf of the Citizen Panel; 25,000 people over the age of 18 were, in both the first and the third wave, randomly drawn from the register. The extracted information was (a) last name, (b) first name, (c) address, (d) gender, (e) age, and (f) phone number (the latter was only included in wave 3). The sample excluded persons without a current home address in Norway. After receiving the data, everyone over the age of 95 was excluded from the sample. Wave 4 of the NCP involved a data collection from existing members of the panel.

3 The CT survey items are based on the work of Karl Dake (1991)—modified by Herron and Jenkins-Smith (2006)—and researchers that have painstakingly translated and adapted CT survey items to Norwegian (see Grendstad et al., 1999; Ollie, 2012).

References

Adam, J.-M. 1992. *Les textes: Types et prototypes*. Paris: Nathan.

Adam, J.-M. 2008. *La linguistique textuelle. Introduction à l'analyse textuelle des discours*. Paris: Armand Colin.

Berinksy, A. J., and D. R. Kinder. 2006. "Making sense of issues through media frames: Understanding the Kosovo crisis." *The Journal of Politics* 68 (3): 640–656.

Berrens, R. P., A. K. Bohara, H. C. Jenkins-Smith, C. L. Silva, and D. L. Weimer. 2003. "The advent of Internet surveys for political research: A comparison of telephone and Internet samples." *Political Analysis* 11: 1–22.

Best, S. J., B. Krueger, C. Hubbard, and A. Smith. 2001. "An assessment of the generalizability of Internet surveys." *Social Science Computer Review* 19: 131–145.

Bruner, J. S. 2003. *Making stories: Law, literature, life*. Cambridge: Harvard University Press.

Campbell, J., and B. Moyers. 2011. *The power of myth*. New York: Anchor.

Dake, K. 1991. "Orienting dispositions in the perception of risk." *Journal of Cross-Cultural Psychology* 22: 61–82.

Douglas, M., and A. Wildavsky. 1982. *Risk and culture: An essay on the selection of technological and environmental dangers*. Berkeley: University of California Press.

Fløttum, K. 2013. "Narratives in reports about climate change." In: Gotti, M., and C. S. Guinda (eds.). *Narratives in academic and professional genres*, pp. 277–292. Bern: Peter Lang.

Fløttum, K., and Ø. Gjerstad. 2013a. "Arguing for climate policy through the linguistic construction of narratives and voices: The case of the South-African green paper 'National Climate Change Response'." *Climatic Change* 118 (2): 417–430.

Fløttum, K., and Ø. Gjerstad. 2013b. "The role of social justice and poverty in South Africa's national climate change response white paper." *South African Journal on Human Rights* 1: 61–90.

Fløttum, K., and Ø. Gjerstad. 2016. "Narratives in climate change discourse." *WIREs Climate Change*. doi: 10.1002/wcc.429

Green, M., and T. C. Brock. 2005. "Persuasiveness of narratives." In: Brock, T. C., and M. C. Green (eds.). *Persuasion: Psychological insights and perspectives*, pp. 117–142, 2nd ed. London: Sage Publications.

Grendstad, G., L. Jensen, G. H. Kristinsson, L. Sjöberg, and S. Sundback. 1999. *Nordic cultures: Measurement, consequences and comparisons: Study description and frequencies*. Bergen: LOS-senteret.

Herron, K. G., and H. C. Jenkins-Smith. 2006. *Critical masses and critical choices: Evolving public opinion on nuclear weapons, terrorism, and security*. Pittsburgh: University of Pittsburgh Press.

Jones, M. D. 2010. *Heroes and villains: Cultural narratives, mass opinions, and climate change*. Ph.D. dissertation. Norman: University of Oklahoma.

Jones, M. D. 2011. "Leading the way to compromise? Cultural theory and climate change opinion." *PS: Political Science & Politics* 44 (4): 720–725.

Jones, M. D. 2014a. "Cultural characters and climate change: How heroes shape our perceptions of climate science." *Social Science Quarterly* 95 (1): 1–39.

Jones, M. D. 2014b. "Communicating climate change: Are stories better than 'just the facts?'" *Policy Studies Journal* 42 (4): 644–673.

Jones, M. D., and M. K. McBeth. 2010. "A narrative policy framework: Clear enough to be wrong?" *Policy Studies Journal* 38 (2): 329–353.

Jones, M. D., M. K. McBeth, and E. A. Shanahan. 2014. "Introducing the narrative policy framework." In: Jones, M. D., E. A. Shanahan, and M. K. McBeth (eds.). *The science of stories: Applications of the narrative policy framework in public policy analysis*, pp. 1–25. Basingstoke & New York: Palgrave MacMillan.

Jones, M. D., and G. Song. 2014. "Making sense of climate change: How story frames shape cognition." *Political Psychology* 35(4): 447–595.

Mattila, A. S. 2000. "The role of narratives in the advertising of experiential services." *Journal of Service Research* 3 (1): 35–45.

Mattila, A. S. 2002. "The use of narrative appeals in promoting restaurant experiences." *Journal of Hospitality & Tourism Research* 26 (4): 379–395.

McBeth, M., M. D. Jones, and E. Shanahan. 2014. "The narrative policy framework." In: Weible, C., and P. Sabatier (eds.). *Theories of the policy process*, pp. 225–266. 3rd ed. Boulder: Westview Press.

Olli, E. D. 2012. *Rejected cultural biases shape our political views*. Ph.D. dissertation. Bergen: University of Bergen.

Polkinghorne, D. E. 1988. *Narrative knowing and the human sciences*. New York: SUNY Press.

Propp, V. 1958. *Morphology of the folktale*. Austin: University of Texas Press.

Rayner, S., and E. Malone. 1998. *The societal framework*. Columbus: Battelle Press.

Thompson, M., R. Ellis, and A. Wildavsky. 1990. *Cultural theory*. Boulder: Westview Press.

Troiani, V., A. Sherry, J. Reilly, and M. Grossman. 2006. "The neural correlates of narrative discourse: An investigation using arterial spin-labeling." *Brain and Language* 99 (1–2):191–192.
Van Dijk, T. A. 1980. *Macrostructures*. Hillsdale, NJ: Lawrence Erlbaum.
Verweij, M., M. Douglas, R. Ellis, Chr. Engel, F. Hendriks, S. Lohmann, et al. 2006. "Clumsy solutions for a complex world: The case of climate change." *Public Administration* 84 (4): 817–843.
Werlich, E. 1976. *A text grammar of English*. Heidelberg: Quelle and Meyer.
Young, K., and J. L. Saver. 2001. "The neurology of narrative." *SubStance* 30 (1/2): 72–84.

Appendix

Instruction to survey participants:

> We would like to know how people understand and interpret texts about the climate question. Below you will see an example of a text on this. We ask you to read the text and afterwards answer some questions. We want to know about the impression you get from the content of the text, and we understand that you will not remember all the details.

Individualistic Narrative

If the world continues with greenhouse gas emissions like today, changes in climate for Norway will include extreme increases in rainfall and dramatic sea level rises. This will constitute a major threat to people and their way of life across the country. The organization Norwegian Business fights climate change by means of a quota trading system (Cap & Trade). This would let businesses reduce CO_2 *emissions through buying and selling GHG emission allowances. However, the radical environmental organization Green World pushes its own destructive agenda. It battles constantly to undermine the free market. But Norwegian Business will expose undue political influence of environmental lobbying groups. It emphasizes that Cap & Trade removes the need for more emission regulations that threaten growth. With the liberty to invest, businesses can keep the economy growing while reducing* CO_2 *emissions.*

Egalitarian Narrative

If the world continues with greenhouse gas emissions like today, changes in climate for Norway will include extreme increases in rainfall and dramatic sea level rises. This will constitute a major threat to people and their way of life across the country. The environmental organization Green World fights climate change by introducing a tax on CO_2*, in order to reduce* CO_2 *emissions. However, the profit-seeking corporate-interest organization Norwegian Business pushes its own greedy agenda. It battles constantly solutions like the* CO_2 *tax in order to line its own pockets. But Green World will expose the undue political influence of the fossil fuel industry lobbying groups. It emphasizes that the* CO_2 *tax will remove economic benefits from climate-hostile industries. With the extra income from this tax, society can undertake a more just distribution of wealth and support the renewable energy sector.*

Metaphors

5 Metaphors in Online Editorials and Op-Eds about Climate Change, 2006–2013

A Study of Germany, the United Kingdom, and the United States

Dimitrinka Atanasova and Nelya Koteyko

Introduction

Print and online media are widely acknowledged and examined as key sources shaping public understanding of climate science and increasingly the political, economic, and social implications of climate change. Indeed, climate change "has more potency now as a mobilizing idea than it does as a physical phenomenon" (Hulme 2009, 328), and it has become one of the most politicized scientific issues attracting abundant media coverage (Deming 2005). As Hoffman writes (2012), the public debate around climate change is no longer about science—"it's about values, culture, and ideology." This development is aptly described by Nisbet (2009), who observes that there are now "two Americas divided along ideological lines"—a Republican America (which questions the validity of climate science and dismisses the urgency of the problem) and a Democratic America (which accepts climate science and is concerned about the issue). This political polarization suggests that climate change has joined the ranks of issues such as taxes and gun control that "define what it means to be a Republican or Democrat" (Nisbet 2009) or even liberal or conservative in a European context (Stokes 2015).

Since the transition of climate change from the domain of science to that of politics in the 1980s (Jaspal and Nerlich 2012), different rhetorical means have been employed to legitimate normative claims about its impacts. Among these, metaphors have attracted significant scholarly attention, particularly as they played a prominent role in efforts to (re-)conceptualize climate change mitigation and adaptation strategies (strategies to reduce greenhouse gas emissions or to prepare for coming changes in the climate, respectively) (Koteyko, Thelwall, and Nerlich 2010; Nerlich and Koteyko 2009; Wise *et al.* 2014). Just as metaphors studied in political communication (Charteris-Black 2006, 2014; Musolff 2016), metaphors employed to contest different policy initiatives and proposed solutions to the climate change problem have high public resonance (Russill 2011) and require critical scrutiny (Atanasova and Koteyko 2015).

In this chapter, we discuss the most frequently used climate-change-related metaphors in editorials and op-eds from three countries that are key to the climate change negotiations (van der Wurff 2009). Similar to another form of persuasive writing—political speeches—editorials/op-eds are intended not only to disseminate information but also to influence beliefs and attitudes (van Dijk 1996). Newspapers often endorse candidates in upcoming elections in editorials, and, thus, they can be seen (together with op-eds) as a form of writing that clearly communicates the kinds of values, cultural beliefs, and ideologies increasingly relevant to climate change understanding. This chapter therefore contributes to the overall goal of establishing "whether metaphor use can reflect or perhaps influence" how climate change is responded to (Musolff 2000, 217), while also addressing calls for more comparative research on climate-change-related metaphors (Koteyko and Atanasova Forthcoming).

Metaphors and Climate Change

The rhetorical role of metaphors is acknowledged by both science communication researchers (Condit, Lynch, and Winderman 2012; Nerlich 2009) and political science scholars (Charteris-Black 2004; Chilton 1996, 2004; Goatly 2007; Musolff 2004). Early studies under the umbrella of science communication identify the wide use of Greenhouse Effect[1] metaphors to simplify and explain how temperature change works (Romaine 1997). Use of Greenhouse Effect metaphors has however declined following the release of the first assessment report of the Intergovernmental Panel on Climate Change (IPCC) in 1990 and the first international Earth Summit in 1992 with which the focus moved from scientific consensus on climate change to the need for policy consensus on mitigation and adaptation efforts (Nerlich 2013).

With this change in focus, research also reports an increased use of War metaphors to communicate the urgency of climate change (Asplund 2010; Cohen 2011; Romaine 1997), which should not be surprising given that War metaphors are seen as potent means of highlighting the seriousness of an issue (Nerlich 2009). And, yet, as already the earliest research on War metaphors in climate change communication has shown, they can also be employed to highlight the divisiveness of the climate change "battle." Romaine (1997, 178), in particular, concludes that, in discussions at the 1992 Earth Summit and its media coverage, War metaphors were used to conceptualize discussions about global warming in terms of a liberal left attacking a conservative worldview—"[environmentalists'] primary enemy: capitalism. [. . .] Their appeals and their scare tactics are designed to transform people into foot soldiers in the army of doomsday environmentalism."

The growing focus on mitigation and adaptation strategies has also given rise to metaphors of Body/Patient and Health/Illness as well as Race/

Movement (Koteyko and Atanasova Forthcoming). Body/Patient metaphors have been employed to express supportive views on geoengineering by likening the planet to a human body and geoengineering to a protective sunscreen or by describing geoengineering as a way of treating the planet and, thus, conceptualizing the planet as a patient and geoengineering as a fitting medicine (Nerlich and Jaspal 2012). In contrast, metaphors of Health/Illness have been used to delegitimize geoengineering as "atmospheric liposuction: a retrospective fix for planetary over-indulgence" (Luokkanen, Huttunen, and Hildén 2014, 976). Finally, metaphors of Race/Movement have been employed in arguments to support carbon capture and storage by conceptualizing it as "a step forward" or "a step in the right direction" (Nerlich and Jaspal 2013, 42) and by talking about a "race for low carbon technologies" and "moving to a low carbon economy" (Nerlich 2012, 40).

Discourse Metaphors

In contrast to early research by proponents of conceptual metaphor theory (Lakoff and Johnson 1980), a recent body of work has focused on both the cognitive and the pragmatic dimensions of metaphor use by investigating how the meanings of conceptually grounded metaphors are shaped by specific sociopolitical contexts. Zinken, Hellsten, and Nerlich (2008) speak of discourse metaphors that are grounded in conceptual metaphors such as KNOWING IS SEEING and AFFECTION IS WARMTH, but whose meaning is shaped by their use at a given time and in the context of a specific topic (for a discussion of conceptual metaphors, see also Chapter 6). This emphasis on situatedness and context in discourse metaphors is particularly important in the comparative study of increasingly contested representations of climate change in the media (Zinken, Hellsten, and Nerlich 2008). As Zinken, Hellsten, and Nerlich (2008, 368) explain, discourse metaphors reflect cultural preoccupations and occur across languages, and they allow for new topics to be discussed "in terms of cultural and mythical commonplaces" and may "tie up with and reinforce long traditions of political thought, ideologies or entrenched cultural values."

Whereas the above-reviewed studies have discussed the role of metaphors in communicating climate science and policy, the political and persuasive dimensions of such communication have received limited attention (Atanasova and Koteyko 2015). In contrast, adopting a standpoint that analogies in public discourse use stereotypical representations of everyday situations to provide evaluative perspectives on contested topics (Musolff 2006) and to legitimize a course of action (Charteris-Black 2006), we examine how metaphors have been used in leading newspapers from Germany, the United Kingdom, and the United States. We ask:

RQ1: What are the most frequently used metaphors in each newspaper?
RQ2: Is the same metaphor recruited in the same way across newspapers?

Sample

We analyzed climate-change-related opinion-page content from three countries that have adopted very different positions and have favored distinct policy options for addressing climate change. Since the first major attempt to develop international climate change policies—the 1992 United Nations Framework Convention for Climate Change (UNFCCC)—Germany has been a proponent of ambitious, binding greenhouse gas emissions reduction targets; the United States has supported less vigorous, nonbinding targets; and the UK has occupied an in-between position (van der Wurff 2009). Over time, all three countries have shown "a remarkable consistency" in adhering to these early stances leading to the suggestion that persistent differences cannot be solely explained with variance in the economic costs and benefits of climate change, but factors such as culture and language may also deserve attention (van der Wurff 2009, 462).

We focused on *Sueddeutsche.de*, *theguardian.com*, and *NYTimes.com*—newspapers that consistently attract vast (and similar) online readerships (Newman and Levy 2014). They also represent comparable political stances, reporting styles, and positions on climate change and can be broadly classified as broadsheet/serious, left-of-center/liberal newspapers of record supportive of mainstream scientific consensus on climate change (Brossard, Shanahan, and McComas 2004; Schäfer, Ivanova, and Schmidt 2014; Woods, Fernández, and Coen 2012). We further concentrated on editorials and op-eds, which, in contrast to news articles, are openly persuasive (Lee and Lin 2006) and, as such, constitute sites where the ideological stances of newspapers are most clearly expressed (Hackett and Zhao 1994). Both editorials and op-eds are examples of opinion journalism (Golan and Munno 2014). Editorials offer a platform for editorial board members to comment on salient issues, while op-eds allow experts not affiliated with the publication to articulate their viewpoints. Despite the prominent function of editorials/op-eds "in the expression and construction of public opinion" (van Dijk 1996) and their continued potential to influence policy makers and the public (Sommer and Maycroft 2008), they remain among the less studied types of ideologically loaded discourse (Day and Golan 2005; Golan and Munno 2014; Parks and Takahashi 2016).

We identified relevant editorials/op-eds by searching the newspaper websites with "climate change" (as an exact phrase) in *theguardian.com* and *NYTimes.com* and "Klimawandel" in *Sueddeutsche.de*. We manually collated editorials/op-eds published between January 1, 2006, and September 30, 2013. These cutoff dates represent important developments—2006 has been identified as the year marking increasing contestation of climate change solutions in print and online news (Koteyko, Thelwall, and Nerlich 2010), and 2013 (September 27 in particular) has marked the release of summaries for policy makers of the Fifth Assessment Report (AR5) of the United Nations Intergovernmental Panel on Climate Change (IPCC) (McGrath 2013). The assessment report itself, which was finalized in 2014, is the most recent in the series of such documents. A total of 527 editorials/op-eds met our search criteria and maintained a focus on climate change (75 in *Sueddeutsche.de*, 191 in *theguardian.com*, 261 in *NYTimes.com*). Such focused content is where the heaviest use of climate-change-related metaphors can be expected to occur (Wallis and Nerlich 2005).

Method

We analyzed editorials/op-eds following the three steps of Charteris-Black's (2004) critical metaphor analysis approach—identification, interpretation, and explanation. At the identification stage, we scrutinized editorials/op-eds for keywords and candidate metaphors, referring to whole linguistic expressions that are metaphorically used. The boundary between metaphorical meanings and the literal meanings of words and expressions can be rather fuzzy (Goatly 1997; Neuman *et al.* 2013; Semino, Heywood, and Short 2004). When deciding if a word/expression was used metaphorically, we therefore considered whether it has a more basic meaning in other contexts, and we consulted the *Macmillan English Dictionary for Advanced Learners* (Rundell 2002) to support our intuition about basic meanings (for a detailed description of the metaphor identification procedure, see Pragglejaz Group 2007).

We systematically lifted the sections containing a metaphor keyword or a candidate metaphor from each editorial/op-ed and entered them in a spreadsheet—the rows of which represented the editorials/op-eds, and the columns recorded the instances (and frequencies) of different metaphor keywords and candidate metaphors (for a list of metaphor keywords, see Table 5.1). In this study, we manually identified metaphors and recorded frequencies, but much progress has also been made in the development of automated metaphor identification and analysis (see Koller *et al.* 2008; Mason 2004; Neuman *et al.* 2013).

Table 5.1 Metaphor Keywords, by Domain

Domain	*Metaphor Keyword*
Illness	causes of the disease; diagnosis; disease; emergency surgery; fever; heart; heart failure; ill; medication; patient; pills; remedies; resuscitation; swallowing pills; symptoms; therapy; treatment
War	attach a stick of dynamite; battle; battle cry; combat; devastation; fight; flag-waver; ravage; retreat; side with; settlement; struggle; surrender; threat; triumph; troops; war; win the battle
Journey	(charting) a course; brakes; backsliding; cross; distance; driving; (heading in the right/wrong) direction; (step) forward; journey; move; path; (on the) road; sail; start; steer; (on the) way

At the second (interpretation) step, we read the recorded sections of text to identify "conceptual metaphors" and "conceptual keys" (Charteris-Black 2004, 38)—that is, we linked metaphorical expressions (candidate metaphors) to the conceptual metaphors they represent. As part of the final (explanation) step, we elaborated on the strategic and persuasive dimension of metaphor use by establishing (1) whether metaphors were used to criticize or endorse pro- or anti-climate-change arguments and (2) what specific functions they fulfilled (for example, support or discredit a mitigation or adaptation strategy). Our working definition of a pro-climate-change argument was an argument that treats the science establishing climate change as settled and/or acknowledges the need to act on it, whereas an anti-climate-change argument was defined as one that questions the above elements (see also Atanasova and Koteyko 2015).

Findings

We identified climate-change-related metaphors in 266 (50.5%) of the total of 527 analyzed editorials/op-eds. We also observed cross-country differences with regard to the most frequently used metaphors (see Table 5.2)—Illness in Germany, War in the UK, and Journey in the United States (RQ1). Differences additionally emerged with respect to the functions that the same metaphor performed across countries (RQ2).

Illness Metaphors in Sueddeutsche.de

In *Sueddeutsche.de*, Illness metaphors were used to advance pro-climate-change arguments and to fulfill functions that can be summarized as (1) efforts to communicate the irrationality of inaction and (2) efforts to conceptualize the state of mitigation and adaptation strategies.

When communicating the danger and irrationality of inaction, opinion-page content depicted the world as a patient—a common, but, as yet, relatively new use of Illness metaphors (Nerlich and Jaspal 2012). An item titled "The world can still be saved" declared that "[t]he symptoms are obvious, the causes of the disease are known—and remedies would be available," and, yet, "the right medication is, apparently, too bitter to swallow. It would consist of a massive shift away from fossil fuels. A therapy which scares industrialized nations far more than the consequences of global warming" (*Sueddeutsche.de*, 06/20/12).

Table 5.2 Frequency of Metaphor Use, by Newspaper

Newspaper	*Total Metaphors*	*Illness*	*War*	*Journey*
Sueddeutsche.de	28.0%	23.8%	4.0%	4.0%
theguardian.com	53.9%	2.1%	43.7%	4.0%
NYTimes.com	54.4%	0%	16.1%	23.7%

The judgmental tone of this statement is consistent with the affordances of Illness metaphors—their bipolar structure (good versus bad) allows for the vivid construction of what is rational and good versus what is not (Arrese and Vara-Miguel 2016). Recalling Sontag's (1989) observations that Illness is a metaphor whose strong moral overtones can dominate individuals' experiences, the absurdity of inaction and the moral overtones of the Illness metaphor were especially powerfully communicated in the following examples:

> The patient is dangerously ill, but to him the therapy seems, although he is not lacking in resources, too expensive. He has anxiously consulted the best specialists; they have all arrived at the same diagnosis. And yet, the patient only takes the medicines sporadically, sometimes not at all. [. . .] But which patient is so stupid? It is the whole world.
>
> (*Sueddeutsche.de*, 09/25/07)

> The earth has a fever and is in an urgent need of treatment. [. . .] But one cannot overcome a disease by denying it or by invoking abstruse theories to prove why the obvious diagnosis may be wrong. Every human in their right mind would try to combat an emerging disease. [. . .] Sure, in theory it is possible that the climate research is wrong and there are other causes to global warming than emissions caused by man. But the one who sees this as a ground for inaction behaves like a patient who has a sore throat, receives a streptococci diagnosis by their doctor, but then comes up with all sorts of theories to disprove the diagnosis—instead of taking precautions and swallowing the pills.
>
> (*Sueddeutsche.de*, 02/06/07)

Rather than legitimize or delegitimize specific solutions as the appropriate cure for climate change (Nerlich and Jaspal 2012) or as a bogus medicine (Luokkanen, Huttunen, and Hildén 2014), *Sueddeutsche.de* opinion-page content employed the Illness metaphor to conceptualize the state of climate change mitigation policies such as the European Union Emissions Trading Scheme (ETS). The ETS, which aims "to commit companies to emit less greenhouse gases [. . . ,] is the heart of European climate policy" but "is now suffering from heart failure due to massive construction errors. For months Europe was debating resuscitation attempts [. . .] emergency surgery failed in the European Parliament" (*Sueddeutsche.de*, 05/20/09). This latter example represents a shift from conceptualizing the world and the earth as a patient, to representing a policy as an ailing body. Such use of the Illness metaphor where political entities are referred to in terms of body, body organs, or functions is an example of rather "clichéd" use (Musolff 2010, 1). As Musolff (2010, 4) argues, to depict a political entity—whether it is a state, a nation, or a policy—as a body is a metaphoric framing that has "a long and famous pedigree in the history of ideas" going back to pre-Socratic thinking, and it constitutes a regular aspect of political discourse.

War Metaphors in theguardian.com

In *theguardian.com*, War metaphors were used to advance pro-climate-change arguments and to fulfill purposes that could be broadly divided into (1) efforts to communicate the urgency to act on climate change and (2) efforts to conceptualize the nature of climate change politics.

Elaborating the unified efforts needed for the prevention of climate change, opinion-page content used war-inspired words such as "danger," "threat," and "fights" (see Bergh 2011; Caruso 2014), which is in line with the common use of War metaphors to emphasize the gravity of the problem and the importance of making an urgent collective effort to solve it (Semino 2008). An item thus stated that "climate change is undeniably a serious threat, and our comments should not be seized upon as an excuse for delay or inaction," but rather "the potential impacts of climate change on the Amazon forest must be a call to action to conserve the Amazon, not a reason to retreat in despair" (*theguardian.com*, 04/07/09). Also emphasizing unified struggle, the work of climate change activists was characterized in terms of war and joint efforts—"unlike many of the fights carried on by their predecessors, today's environmental protesters find themselves pushing at an open door. [. . .] [T]he British government, and Tony Blair especially, may talk a better game in tackling climate change than its final actions reveal, but no one can seriously accuse it of not appreciating the danger" (*theguardian.com*, 09/01/06). The moral undertones inherent in War metaphors further allowed for negative evaluations of those failing to join the "fight." An item declaring that "the US is late to the climate-change fight" argued that "in the absence of a progressive strategy, mere tactics fill the void" (*theguardian.com*, 06/26/09).

War metaphors were also employed to conceptualize the nature of climate change politics, which indicates that the fight against climate change is not only a unifying force but also a divisive force. Reflecting the politicization of climate change, such metaphors alert us that the conditions and participants in the metaphorical battle of climate change have been radically altered and "us against climate change" has become "us against them." An item provoked by Heartland Institute's billboard campaign to show that prominent advocates of global warming are not scientists spoke of "the forces of climate change denial" that have "fought with remarkable tenacity to stall and delay the inevitable recognition that we're in serious trouble," yet "eventually, climate denial will 'lose', because physics and chemistry are not intimidated, even by Lord Monckton," but "if he and his ilk, a crew of certified planet wreckers, delay action [. . .] they'll be able to claim one of the epic victories in political history—one that will last for geological epochs" (*theguardian.com*, 06/04/12). Such instances follow the longstanding use of War metaphors in political communication (Charteris-Black 2006) and render climate change part and parcel of politics as a form of warfare (Larson, Nerlich, and Wallis 2005): "Malcolm Turnbull, leader of the conservative coalition in which the Liberals are the dominant force, ordered his troops to support the government's plans to combat climate change" (*theguardian.com*, 08/28/11).

Journey Metaphors in NYTimes.com

In *NYTimes.com*, Journey metaphors were used to advance pro-climate-change arguments and to fulfill purposes such as (1) conceptualizing efforts to address climate change and (2) supporting or denouncing solutions.

When conceptualizing efforts to address climate change, the focus was on the journeying—taking steps and moving—not on the destination. In an item titled "An important step on global warming," President Obama's proposed federal limits on power plant emissions of carbon dioxide were described as "a move that has already caused dismay" and "a welcome sign of his determination to move ahead in his own authority and bypass Congress" (*NYTimes.com*, 09/22/13). Another item reflected on how "[o]fficials from the Obama administration have been beating a steady path to China's door to talk about climate change," and, yet, China has not committed to specific measures. As a result, neither has the United States—"a House committee bill that seeks a mandatory 80 percent cut in America's emissions by mid-century has barely begun its journey through Congress" (*NYTimes.com*, 06/15/09). This pronounced use of Journey metaphors, especially in relation to President Obama, in fact overlaps with his communication style. Journey has been identified as Obama's "governing metaphor" (Cox 2012, 1), of which he has made extensive use ever since his 2004 speech at the Democratic National Convention.

The Journey metaphor was also invoked indirectly through other entailments such as movement down a path or pointing the way. We were, for example, told that California, "the first state to mandate reductions of greenhouse gas emissions from every sector of the economy," has "pointed the way"—"a hard act to follow for any governor, but to some extent Mr. Spitzer's path has already been smoothed by his predecessor" (*NYTimes.com*, 02/20/07). This emphasis on journeying—which helps to construct US efforts in a positive light by focusing on the process, movement, and hardships along the way to suggest progress—mimics the strategies of business communication on the related issue of sustainability (Ihlen and Roper 2014; Milne, Kearins, and Walton 2006).

Journey metaphors were further used to support or denounce solutions to climate change as a move in the right or wrong direction (see also Nerlich and Jaspal 2013). An item stated that "the surest way to address the threats of climate change is to put a price on the carbon emissions that power plants, cars and other users of fossil fuels emit." President Obama, who "has now stepped forward" with measures along these lines, was lauded for having made "a worthy start" (*NYTimes.com*, 06/25/13). Another item (which proclaimed that "[w]e need sustainable clean-tech jobs [. . .] [a]nd we need to take steps to mitigate climate change") identified a price on carbon as "the easiest way to do all of this at once" and "start taking steps" (*NYTimes.com*, 09/13/11). Other solutions such as Keystone XL, an oil pipeline system in Canada and the United States commissioned in 2010, were however denounced—"if the public were to allow our well-oiled government to shepherd it to existence," this would be "the first step down the wrong road, perpetuating our addiction to dirty fossil fuels, moving to even dirtier ones" (*NYTimes.com*, 03/04/13).

Illness and Journey Metaphors: Different Functions across Newspapers

In contrast to *Sueddeutsche.de*, *NYTimes.com* did not employ the Illness metaphor, and *theguardian.com* editorials/op-eds used it only to support certain solutions to climate change. With regard to geoengineering, we were told that "there is no pain-free fix—climate change will force fundamental changes in the way we live. But if easy panaceas are a delusion, so too is any hope that cutting consumption, without exploiting science, can do the trick alone" (*theguardian.com*, 07/13/09).

In *Sueddeutsche.de* and *theguardian.com*, unlike in *NYTimes.com*, the Journey metaphor was used exclusively to conceptualize efforts to address climate change. A *Sueddeutsche.de* item that discussed President Obama's plans to impose stringent climate targets on the American auto industry praised him for having made "many small steps that go in the right direction" and for demonstrating that "[t]he steamer America is beginning to change course—too late, too slow, but still it is a start. [. . .] He must convince his country that in the long run this can be but the winning path" (*Sueddeutsche.de*, 08/07/12). Similarly, an item from *theguardian.com* that reflected on the prospects of "a world beyond oil" stated that "the country is on course to miss its target of a 30% cut in emissions by 2020," which is why "[n]ext week's climate change bill [. . .] must be strengthened by parliament before it becomes law in 2008. Dreaming of a future beyond the oil age is easy. Getting there will be painful" (*theguardian.com*, 03/06/07). Another *theguardian.com* item on the outcomes of the 2007 summit of the G8 forum announced that the deal "is not even the beginning of the end of the quest for a global framework to tackle climate change. But world leaders did at last agree a direction of travel" (*theguardian.com*, 06/08/07).

War Metaphors: Same Functions across Newspapers

In all three newspapers, the War metaphor was used to communicate the urgency to act on climate change and to conceptualize the nature of climate change politics. What is interesting is the absence of the word "war" or "Krieg" from lexicalizations of the War metaphor in *Sueddeutsche.de*, for which several explanations are possible. First, there appears to be a preference in German to use "Kampf" (meaning "fight") to denote metaphorical war and to use "Krieg" (meaning "war") to denote real warfare (Barrett 2007). Second, "Krieg" has acquired rather negative connotations post-1945, particularly associations with total destruction and the Third Reich (Bréal 1991).

As an example of using the War metaphor to communicate the urgency to act on climate change, a *Sueddeutsche.de* item dealing with the Cancun Climate Change Conference stated that "[n]othing is won yet in the fight for climate protection, as the Cancun summit left many questions unanswered." Cancun laid the foundations "for a financial fund to support poor countries in their future fight against climate change," but this is not enough: "So all is well with the climate then? Not at all. [. . .] China and the US want to continue fighting global warming by using measures that are not legally binding" (*Sueddeutsche.de*, 02/19/13). Similarly, a *NYTimes.com* item lamented that "the effort to combat global warming has flagged as other crises have commanded the attention of politicians and the public," and, yet, while climate change impacts are "difficult to predict, society would be wise to move swiftly and aggressively to minimize them" (*NYTimes.com*, 05/23/10).

As an example of the use of war language to conceptualize the nature of climate change politics, a *Sueddeutsche.de* op-ed recounting the outcomes of the Copenhagen Climate Change Conference described it as "a great defeat" for "the climate chancellor Angela Merkel"—"she unusually candidly explains the extent of the debacle: 'There is only a general consensus to limit global warming to two degrees, but no obligation.'" The conclusion is that "the German climate chancellor, opinion leader in the European Union, internationally acclaimed negotiator, she did not make it" (*Sueddeutsche.de*, 06/21/12), suggesting that climate politics is a fight in which there are winners and losers. In *NYTimes.com*, the divisive nature of the climate change "battle" across (but also within) party lines manifested when John McCain, senior Senator from Arizona, was described as someone who "has been engaged in the fight against global warming for years" even though "[e]ndorsing an aggressive and potentially expensive effort to reduce carbon emissions will not win him friends on the right wing" (*NYTimes.com*, 05/14/08). Another *NYTimes.com* item that described a new action plan by President Obama as one that "can bypass the anti-environmentalists who control the House of Representatives" stated that they realize this but "don't seem eager to attack climate science. [. . .] Instead, they're going for the economic angle, denouncing the Obama administration for 'waging a war on coal' that will destroy jobs" (*NYTimes.com*, 06/27/13).

Discussion

We analyzed editorials/op-eds published in leading newspapers from Germany, the UK, and the United States over the span of seven years (2006–2013) in order to answer two principal questions: what metaphors were used most frequently; and whether the same metaphor was recruited in the same way across newspapers. We found that newspapers differed as to the preferred source domain for metaphors and in terms of the functions that the same metaphor fulfilled. The metaphors of Illness, War, and Journey (most prominent in *Sueddeutsche.de*, *theguardian.com*, and *NYTimes.com*, respectively) have a history of research in both environmental and political discourse and speak strongly to German, British, and American national identity and culture. In what follows, we try to further account for their prominent use and draw out implications.

The use of Illness metaphors in environmental discourse dates back to the 1980s, when such metaphors were employed to describe the state of forests in discussions of acid rain and forest decline (Väliverronen 1998). Climate scientists like Somerville (2006, 1) have declared themselves "planetary physicians" and have recommended Body/Patient metaphors as especially effective in conveying the negative consequences of climate change and motivating action. But environmental communication research also suggests that by anchoring climate change in the fear of disease, metaphors of Illness might, in fact, demotivate action (Höijer 2010). The media is said to contribute to a culture of fear with its daily reports on scare stories, and, therefore, frequent exposure to messages conceptualizing climate change as an illness may lead to it becoming part of the expanding sphere of fear appeals in the media, to which the public gradually becomes indifferent (Höijer 2010).

Given our overall tendency to metaphorize phenomena that cause problems (Kövecses 2002), it is unsurprising that climate change is conceptualized as an illness and that means to address it are conceptualized as treatments. The especially prominent use of Illness metaphors in *Sueddeutsche.de* can be tentatively linked to the above-described early uses of such metaphors to discuss acid rain and forest dieback coupled with the prominent place that the forest takes in German culture ("Love for the forests deeply rooted in German psyche" 2009). An emotive campaign in the early 1980s around "Waldsterben" (translated as "forest dieback") is believed to have contributed to the political power of the Green Party in Germany (Newbery 2003). Further, the word "Waldsterben" is more affective in its German version—the emotional charge of the combined use of "forest" ("Wald") and "dying/death" ("sterben") is not captured in the English term that denotes this process.

Illness metaphors are also regarded as politically potent—they can prompt action by appealing to universal human experiences of illness (Nancy *et al.* 1997). Cross-linguistic research into Illness metaphors in political discourse confirms that many issues are described through human ailment and disease leading to the conclusion that such metaphors are deeply entrenched in cognition and reasoning (Urbonaitė and Šeškauskienė 2007). Further, representations of nations, states, and policies as diseased bodies have long roots in European discourse (Musolff 2010), and the transfer of such political discourse metaphors to climate change discourse is a clear manifestation of the politicization of the issue. In a German context, metaphors of Illness (particularly of infectious disease) and the related depictions of countries and nations as ailing bodies are—due to historical reasons—highly problematic (Musolff 2010). As Musolff (2010) notes, a German politician conceptualizing a sociopolitical conflict via Illness/Body metaphors would be quickly discredited. Conceptualizing the world or the earth as a body or patient, as was the case in *Sueddeutsche.de*, does not carry the same sinister connotations and, in fact, taps into a long tradition of corporeal imagery in German political discourse that can be traced back to the sixteenth century (Musolff 2010).

Similarly, War metaphors are regularly used in discussions of climate change (Koteyko and Atanasova Forthcoming), and our findings that War is the most popular metaphor in *theguardian.com* editorials/op-eds resonate with past research of climate change reporting in Anglophone nations. Such research has identified British newspapers as most heavily drawing on War metaphors (Cohen 2011). This salient use of war imagery may be a welcome development—War metaphors could create the necessary conditions for policy makers to advocate ambitious proposals for dealing with climate change, such as making greenhouse gas reduction proposals reminiscent of wartime austerity programs (Cohen 2011). Alternatively, the use of War metaphors to refer to climate change politics and in-group fighting, as was the case in *theguardian.com*, represents a typical use of War metaphors in political communication where politics is conceptualized as a battle, and it has little to do with instigating action on climate change (Atanasova and Koteyko 2015). Thus, War metaphor use may do the opposite and contribute to climate fatigue (Ward 2008), as the entrenched POLITICS IS WAR metaphor has become coded speech, which, with repetition, can lead to "dulling the critical faculties rather than awakening them" (Mio 1997, 119).

In broader terms, some of the most powerful stories in every culture are centered on war and struggle—"waging and winning a war makes a good story and also makes a good frame to tell stories about other events" and issues (Nerlich 2004, 19). Wars have visible endpoints, such as gaining control and winning a battle, which is also what makes them so attractive in political discourse (Nerlich 2004). This is even truer in a UK context, where "war symbolized in the metaphors of war used so widely and so frequently" is a major component of the national character (Mangan 1996, 12). As Duffy (2001, xi) writes, the myth of the English/British (two identity constructs that are still difficult to disentangle) "has always thrived best when we were at loggerheads with Spain, France or Germany and under threat of invasion of our sacred soil, our island fortress." As anyone living in the UK would also have noticed, there has been much focus on commemoration activities in 2014 and 2015 around the two world wars from which the country emerged victorious. Indeed, "as the 20th century drew to its close, artistic, media and scholarly interest in war/s appeared particularly strong in Britain" (Korte 2002, 1).

Employing Journey metaphors in climate change communication is not unusual either (Nerlich 2012; Nerlich and Jaspal 2013). By focusing on taking steps without discussing the end point or destination, as in *NYTimes.com* editorials/op-eds, Journey metaphors seem to close down criticism—after all, things are moving (Ihlen and Roper 2014). There is only a shadow of criticism in the admission that a House committee bill that seeks a mandatory 80% cut in America's emissions by mid-century has just begun its journey through Congress. This resonates with critiques of the Journey metaphor as only appearing to engage with the problem at hand and embrace change while actually reinforcing business-as-usual (Ihlen and Roper 2014; Milne, Kearins, and Walton 2006). Especially in those cases when discussion of the final destination is omitted, Clancy (1999, 42) calls the Journey metaphor "a problematic guide to action."

Yet Journey can be a politically useful metaphor, and, as Milne, Kearins, and Walton (2006) put it, it can offer complexity and hardships along the way as a potential excuse for lack of substantial progress while also expressing a notion of progress, if not actual achievement by virtue of simply embarking on the journey. In one of our examples, China's reluctance to commit to specific greenhouse gas reduction targets is proffered as an explanation for a bill muddling through Congress. With no defined end point or goal—for example, a reduction of greenhouse gas emissions of certain magnitude—it is always possible to show progress vis-à-vis a previous state of affairs (Milne, Kearins, and Walton 2006). The Journey metaphor is also attractive for being an archetypical metaphor that transcends time and culture and resonates at a deep, subconscious level (Cox 2012; Darsey 2009). It can be expected to resound particularly well with the American public, as the notion of a journey very much defines what it means to be American and as the country is often envisioned as a voyaging ship ("Editorial: The American journey" 2009). This expectation could have also motivated President Obama's extensive use of journey language (Cox 2012).

In conclusion, our findings that Illness was the most frequently used metaphor in *Sueddeutsche.de* opinion-page content, War in *theguardian.com*, and Journey in *NYTimes.com* indicate that diverging national efforts to communicate the gravity of anthropogenic climate change are closely intertwined with national contexts and cultures, political identities, and policy making. Such tailoring to national identity and culture can be seen as a positive step to maximize public resonance by appealing to a set of recognizable values and norms. But there also is a danger, for Illness, War, and Journey metaphors are deeply entrenched in political communication, and their transfer to climate change discourse risks turning the issue into "politics as usual"—a development that defies Russill's (2011) famous call for new and creative metaphors as a way to provoke climate change action.

Note

1 Double quote marks denote direct examples; underlining signifies metaphorical expressions within quotes; small capitals indicate conceptual metaphors; italicized small capitals indicate systematic metaphors; and initial capitals label metaphors more generally.

References

Arrese, Á., and A. Vara-Miguel. 2016. "A comparative study of metaphors in press reporting of the Euro crisis." *Discourse & Society* 27 (2): 133–155.

Asplund, M. 2010. "Metaphors in climate discourse: An analysis of Swedish farm magazines." *Journal of Science Communication* 10 (4): 1–8.

Atanasova, D., and N. Koteyko. 2015. "Metaphors in *Guardian Online* and *Mail Online* opinion-page content on climate change: War, religion, and politics." *Environmental Communication*: 1–18. doi: 10.1080/17524032.2015.1024705

Barrett, D. 2007. " 'War on terror'—An intentional choice of words? A corpus analysis of war on and war against." Paper presented at the Corpus Linguistic Conference, University of Birmingham, Birmingham, July 27–30.

Bergh, G. 2011. "Football is war: A case study of minute-by-minute football commentary." *Veredas on Line—Temática* 2: 83–93.

Bréal, M. 1991. *The beginnings of semantics: Essays, lectures and reviews*. Stanford: Stanford University Press.

Brossard, D., J. Shanahan, and K. McComas. 2004. "Are issue-cycles culturally constructed? A comparison of French and American coverage of global climate change." *Mass Communication and Society* 7 (3): 359–377.

Caruso, A. 2014. "A roadmap to peace: Journey metaphors in political speeches on the Middle East peace process." *Lingue e Linguaggi* 11: 77–92.

Charteris-Black, J. 2004. *Corpus approaches to critical metaphor analysis*. Basingstoke & New York: Palgrave MacMillan.

Charteris-Black, J. 2006. *Politicians and rhetoric: The persuasive power of metaphor*. Basingstoke & New York: Palgrave MacMillan.

Charteris-Black, J. 2014. *Analysing political speeches: Rhetoric, discourse and metaphor*. Basingstoke & New York: Palgrave MacMillan.

Chilton, P. 1996. *Security metaphors: Cold War discourse from containment to common European home*. Peter Lang: Berne and New York.

Chilton, P. 2004. *Analysing political discourse: Theory and practice*. London: Routledge.

Clancy, J. J. 1999. *The invisible powers: The language of business*. Lexington: Lexington Books.

Cohen, M. J. 2011. "Is the UK preparing for 'war'? Military metaphors, personal carbon allowances, and consumption rationing in historical perspective." *Climatic Change* 104: 199–222.

Condit, C. M., J. Lynch, and E. Winderman. 2012. "Recent rhetorical studies in public understanding of science: Multiple purposes and strengths." *Public Understanding of Science* 21 (4): 386–400.

Cox, J. L. 2012. "Politics in motion: Barack Obama's use of movement metaphors." *American Communication Journal* 14 (2): 1–13.

Darsey, J. 2009. "Barack Obama and America's journey." *Southern Communication Journal* 74 (1): 88–103.

Day, A., and G. Golan. 2005. "Source and content diversity in op-ed pages: Assessing editorial strategies in *The New York Times* and *The Washington Post*." *Journalism Studies* 6 (1): 61–71.

Deming, D. 2005. "Global warming, the politicization of science, and Michael Crichton's state of fear." *Journal of Scientific Exploration* 19 (2): 247–256.

Duffy, M. 2001. *England: The making of the myth*. London: Fourth Estate.

Editorial: The American journey. 2009. *The Washington Times*, October 12.

Goatly, A. 1997. *The language of metaphors*. London: Routledge.

Goatly, A. 2007. *Washing the brain—metaphor and hidden ideology*. Amsterdam & Philadelphia: John Benjamins Publishing Company.

Golan, G. J., and G. Munno. 2014. "Few Latin American items appear on editorial pages." *Newspaper Research Journal* 35 (1): 20–35.

Hackett, R. A., and Y. Z. Zhao. 1994. "Challenging a master narrative: Peace protest and opinion editorial discourse in the United-States press during the Gulf War." *Discourse & Society* 5 (4): 509–541.

Höijer, B. 2010 "Emotional anchoring and objectification in the media reporting on climate change." *Public Understanding of Science* 19 (6): 717–731.

Hoffman, A. J. August 17 2012. "Climate science as culture war." *Stanford Social Innovation Review*: 31–37.

Hulme, M. 2009. *Why we disagree about climate change: Understanding controversy, inaction and opportunity*. Cambridge: Cambridge University Press.

Ihlen, Ø., and J. Roper. 2014. "Corporate reports on sustainability and sustainable development: 'We have arrived'." *Sustainable Development* 22: 42–51.

Jaspal, R., and B. Nerlich. 2012. "When climate science became climate politics: British media representations of climate change in 1988." *Public Understanding of Science* 23 (2): 122–141.

Koller, V., A. Hardie, P. Rayson, and E. Semino. 2008. "Using a semantic annotation tool for the analysis of metaphor in discourse." *metaphorik.de* 15: 141–160.

Korte, B. 2002. "Wars and 'British' identities: From Norman conquerors to Bosnian warriors." In: Korte, B., and R. Schneider (eds.). *War and the cultural construction of identities in Britain*, pp. 9–24. Amsterdam: Rodopi.

Koteyko, N., and D. Atanasova. 2017. "Metaphor and the representation of scientific issues: Climate change in online and print media." In: Semino, E., and Z. Demjén (eds.). *The Routledge handbook of metaphor and language*. London: Routledge.

Koteyko, N., M. Thelwall, and B. Nerlich. 2010. "From carbon markets to carbon morality: Creative compounds as framing devices in online discourses on climate change mitigation." *Science Communication* 32 (1): 25–54.

Kövecses, Z. 2002. *Metaphor: A practical introduction*. Oxford & New York: Oxford University Press.

Lakoff, G., and M. Johnson. 1980. *Metaphors we live by*. Chicago: Chicago University Press.

Larson, B., B. Nerlich, and P. Wallis. 2005. "Metaphors and biorisks." *Science Communication* 26: 243–268.

Lee, F. L. F., and A. M. Y. Lin. 2006. "Newspaper editorial discourse and the politics of self-censorship in Hong Kong." *Discourse Society* 17 (3): 331–358.

Love for the forests deeply rooted in German psyche. 2009. *Deutsche Welle*, January 23.

Luokkanen, M., S. Huttunen, and M. Hildén. 2014. "Geoengineering, news media and metaphors: Framing the controversial." *Public Understanding of Science* 23 (8): 966–981.

Mangan, J. A. 1996. "Duty unto death: English masculinity in the age of the new imperialism." In: Mangan, J. A. (ed.). *Tribal identities: Nationalism, Europe, sport*, pp. 10–38. London: Frank Cass.

Mason, Z. 2004. "CorMet: A computational, corpus-based conventional metaphor extraction system." *Computational Linguistics* 30 (1): 23–44.

McGrath, M. 2013. "IPCC climate report: Humans 'dominant cause' of warming." *BBC*, September 27.

Milne, M. J., K. Kearins, and S. Walton. 2006. "Creating adventures in Wonderland: The journey metaphor and environmental sustainability." *Organization* 13 (6): 801–839.

Mio, J. S. 1997. "Metaphor and politics." *Metaphor & Symbol* 12 (2): 113–133.

Musolff, A. 2000. "Political imagery of Europe: A house without exit doors?" *Journal of Multilingual and Multicultural Development* 21 (3): 216–229.

Musolff, A. 2004. "Metaphor and conceptual evolution." *metaphorik.de* 7: 55–75.

Musolff, A. 2006. "Metaphor scenarios in public discourse." *Metaphor & Symbol* 2 (1): 23–38.

Musolff, A. 2010. *Metaphor, nation and the Holocaust: The concept of the body politic*. New York: Routledge.

Musolff, A. 2016. *Political metaphor analysis: Discourse and scenarios*. London: Bloomsbury Academic.

Nancy, R., J. Eyles, D. Cole, and A. Iannantuono. 1997. "The ecosystem metaphor in science and policy." *The Canadian Geographer* 41 (2): 114–127.

Nerlich, B. 2004. "War on foot and mouth disease in the UK, 2001: Towards a cultural understanding of agriculture." *Agriculture and Human Values* 21: 15–25.

Nerlich, B. 2009. " 'The post-antibiotic apocalypse' and the 'war on superbugs': Catastrophe discourse in microbiology, its rhetorical form and political function." *Public Understanding of Science* 18 (5): 574–590.

Nerlich, B. 2012. " 'Low carbon' metals, markets and metaphors: The creation of economic expectations about climate change mitigation." *Climatic Change* 110: 31–51.

Nerlich, B. 2013. "Coming to terms with climate change through metaphors." Paper presented at the International Science Forum Heidelberg "Risk discourses, discourse risks: European perspectives on the linguistic depiction of technological risks," Heidelberg, September 9–11.

Nerlich, B., and R. Jaspal. 2012. "Metaphors we die by? Geoengineering, metaphors, and the argument from catastrophe." *Metaphor and Symbol* 27 (2): 131–147.

Nerlich, B., and R. Jaspal. 2013. "UK media representations of carbon capture and storage: Actors, frames and metaphors." *Metaphor and the Social World* 3 (1): 35–53.

Nerlich, B., and N. Koteyko. 2009. "Compounds, creativity and complexity in climate change communication: The case of 'carbon indulgences." *Global Environmental Change* 19: 345–353.

Neuman, Y., D. Assaf, Y. Cohen, M. Last, S. Argamon, N. Howard, and O. Frieder. 2013. "Metaphor identification in large texts corpora." *PLoS ONE* 8 (4): e62343.

Newbery, D. M. 2003. "Sectoral dimensions of sustainable development: Energy and transport." *Economic Survey of Europe* 2: 73–93.

Newman, N., and D. A. L. Levy. 2014. "Reuters Institute Digital News Report 2014." Reuters Institute. https://reutersinstitute.politics.ox.ac.uk/sites/default/files/Reuters%20Institute%20Digital%20News%20Report%202014.pdf

Nisbet, M. 2009. "Communicating climate change: Why frames matter for public engagement." *Environment*, March–April.

Parks, P., and B. Takahashi. 2016. "From apes to whistleblowers: How scientists inform, defend, and excite in newspaper op-eds." *Science Communication* 38 (3): 275–302.

Pragglejaz Group. 2007. "MIP: A method for identifying metaphorically used words in discourse." *Metaphor & Symbol* 22 (1): 1–39.

Romaine, S. 1997. "War and peace in the global greenhouse: Metaphors we die by." *Metaphor & Symbolic Activity* 11: 175–194.

Rundell, M. 2002. *Macmillan English dictionary for advanced learners*. Oxford: Macmillan.

Russill, C. 2011. "Temporal metaphor in abrupt climate change communication: An initial effort at clarification." In: Leal Filho, W. (ed.). *The economic, social, and political elements of climate change*, pp. 113–132. New York: Springer.

Schäfer, M. S., A. Ivanova, and A. Schmidt. 2014. "What drives media attention for climate change? Explaining issue attention in Australian, German and Indian print media from 1996 to 2010." *The International Communication Gazette* 76 (2): 152–176.

Semino, E. 2008. *Metaphor in discourse*. Cambridge: Cambridge University Press.

Semino, E., J. Heywood, and M. Short. 2004. "Methodological problems in the analysis of metaphors in a corpus of conversations about cancer." *Journal of Pragmatics* 36: 1271–1294.

Somerville, R. C. J. 2006. "Medical metaphors for climate issues." *Climatic Change* 76: 1–6.

Sommer, B., and J. R. Maycroft. 2008. "Influencing public policy: An analysis of published op-eds by academics." *Politics & Policy* 36: 586–613.

Sontag, S. 1989. *AIDS and its metaphors*. New York: Farrar, Straus and Giroux.

Stokes, B. 2015. *The U.S. isn't the only nation with big partisan divides on climate change*. Pew Research. http://www.pewresearch.org/fact-tank/2015/11/06/the-u-s-isnt-the-only-nation-with-big-partisan-divides-on-climate-change/

Urbonaitė, J., and I. Šeškauskienė. 2007. "Health metaphor in political and economic discourse: A cross-linguistic analysis." *Studies about Languages* 11: 68–73.

Väliverronen, E. 1998. "Biodiversity and the power of metaphor in environmental discourse." *Science Studies* 11 (1): 19–34.

van der Wurff, R. 2009. "Climate change policy of Germany, UK and USA." In: Boersema, J. J., and L. Reijnders (eds.). *Principles of Environmental Sciences*, pp. 459–471. New York: Springer.

van Dijk, T. A. 1996. "Opinions and ideologies in editorials." Paper presented at the Fourth International Symposium of Critical Discourse Analysis, Language, Social Life and Critical Thought, Athens, December 14–16.

Wallis, P., and B. Nerlich. 2005. "Disease metaphors in new epidemics: The UK media framing of the 2003 SARS epidemic." *Social Science & Medicine* 60: 2629–2639.

Ward, B. 2008. "2008's year-long fall-off in climate coverage; tracking the trends, and the reasons behind them." *The Yale Forum on Climate Change & the Media*, December 12.

Wise, R. M., I. Fazey, M. Stafford Smith, S. E. Park, H. C. Eakin, E. R. Archer van Garderen, and B. Campbel. 2014. "Reconceptualising adaptation to climate change as part of pathways of change and response." *Global Environmental Change* 28: 325–336.

Woods, R., A. Fernández, and S. Coen. 2012. "The use of religious metaphors by UK newspapers to describe and denigrate climate change." *Public Understanding of Science* 21 (3): 323–339.

Zinken, J., I. Hellsten, and B. Nerlich. 2008. "Discourse metaphors." In: Frank, R. M., Dirven, R., T. Ziemke, and E. Bernárdez (eds.). *Body, language and mind: Vol. 2. Sociocultural situatedness*, pp. 363–386. Amsterdam: John Benjamins.

6 Conceptual Metaphors Associated with Climate Change in Corporate Reports in the Fossil Fuels Market

Two Perspectives from the United States and Australia

Oleksandr Kapranov

Introduction

The issue of climate change is a multifaceted and complex phenomenon. The complexity of climate change discourse involves a plethora of variables, which encompass the input from political, nongovernmental, and corporate actors (Fløttum and Dahl 2012). From the vantage point of corporate actors, the issue of climate change involves multiple aspects associated with fossil fuels—for instance, the combustion of fossil fuels resulting in greenhouse gas (GHG) emissions. The negative impact of fossil fuels and GHG on climate change is well known in corporate world (Livesey 2002). Corporate awareness of climate change is manifested in fossil fuels corporations' discourse involving climate change, the protection of the environment, and ecological sustainability (Talbot and Boiral 2014).

Climate change discourse by fossil fuels corporations is reflective of the measures undertaken by corporate actors to mitigate the negative consequences of human-induced climate change. There is a contention in the research literature that by analyzing corporate climate change discourse, non-corporate actors and the public at large can get a better understanding of corporate decision making as well as corporate actions to mitigate negative consequences of climate change (Allen 2016). This chapter will further discuss climate change discourse by two different corporations and provide insight into how these corporate actors present themselves to the public in their business reports concerning the issue of climate change.

The chapter involves climate change discourse by ExxonMobil (United States) and BHP Billiton (Australia), each considered one of the major players in the fossil fuels market in North America and Australia, respectively. Climate-change-related documents by ExxonMobil (further referred to in the chapter as Exxon) and BHP Billiton (further referred to as BHP) will be examined by means of a qualitative study aimed at identifying conceptual metaphors associated with the issue of climate change. The focus of the study is on investigating whether or not there are qualitative differences in conceptual metaphors associated with climate change identified in corporate reports by Exxon and BHP.

The novelty of this research rests with the following two considerations: First, the study involves a juxtaposition of Exxon's and BHP's climate change discourse. While there is previous research that addresses climate change discourse separately by Exxon (Livesey 2002) and by BHP (Nyberg, Spicer, and Wright 2013), currently there is no scientific literature involving a juxtaposition of climate change discourse by these two corporations. Presumably, the juxtaposition will reveal differences or similarities in the environmental practices of these two corporations. Second, the study describes a corporate perspective on climate change by BHP, a leading fossil fuels company in Australia. Given that corporate discourse of climate change in North America seems to be well researched (Livesey 2002), its analogue in the Southern Hemisphere is significantly underrepresented (Czerniewicz, Goodier, and Morrell 2016, 4).

This chapter is structured as follows: First, an outline of previous qualitative studies involving corporate discourse associated with climate change will be presented. Second, previous research involving conceptual metaphors in climate change discourse will be discussed. Third, the present qualitative study will be introduced, aimed at identifying conceptual metaphors associated with the issue of climate change in corporate reports by Exxon and BHP. Fourth, the conceptual metaphors identified in the material from each company will be juxtaposed in order to discuss the Australian and American corporate perspectives on the issue of climate change and mitigation.

Previous Qualitative Studies Involving Australian and US Corporate Discourse Associated with Climate Change

It can be assumed that US corporate discourse involving climate change is qualitatively different from its counterparts in the rest of the Northern Hemisphere (Greenberg, Knight, and Westersund 2011). In particular, this contention holds true for the period between 2000 and 2010, which is marked by the presidency of G.W. Bush (Good 2008, 236–237). Research indicates that climate change discourse by North American fossil fuels corporations is characterized by "inconsistency, ambiguity, heterogeneity, and limited scope" (Jones and Levy 2007, 4). Specifically, the ambiguity of corporate climate change discourse is evident from the substitution of the term 'climate change' for 'global warming', as well as from a skeptical tonality toward climate change adopted by some North American corporate actors (Good 2008).

Climate change discourse by corporate actors in the United States is marked by their resistance to scientific evidence involving negative consequences of global climate change (Pulver 2007). In particular, Exxon has been reported to question the findings of climate scientists. In this regard,

Pulver (2007, 54) indicates that "ExxonMobil best exemplifies an adversarial climate policy." Exxon's skepticism toward climate change feeds into the corporation's climate change discourse, which appears to question climate science and to doubt the validity of scientific research involving GHG emissions. Additionally, climate change discourse by Exxon deemphasizes the role of international cooperation and partnership on the issue of global climate change. Exxon approaches the issue of climate change from a strategic business investment point of view rather than from a public policy perspective (Pulver 2007).

Exxon's skeptical stance on climate change is reported to be present in the corporation's official reports and corporate statements, as well as in so-called advertorials, discourse by Exxon-affiliated lobbying groups, and political discourse by political actors sponsored by Exxon (Pulver 2007, 54). In her seminal study of climate change discourse by Exxon, Livesey (2002) analyzes Exxon's advertorials published in the *New York Times*. Livesey (2002, 126) observes that the advertorials are centered on climate change skeptic topics—e.g., complaints about the anti-competitive and costly to the consumer aspects of environmental regulations, the central role of fossil fuels in economic development in the foreseeable future, the need for more research involving climate due to the supposedly inconclusive scientific evidence, etc.

Livesey (2002, 127) posits that Exxon's climate change discourse is framed by a rhetorical strategy "to turn environmental protection . . . into the enemy, business and technology into saviors, and environmentalists and governments into incompetent meddlers who would do no good." Livesey (2002, 127) suggests that Exxon's climate change discourse varies and involves a concatenation of images, such as the appreciation of technology as a means of averting negative consequences of climate change, a construal of the image of responsibility, and an image of 'win-win environmentalism', which is beneficial both for nature and for corporate profit (Livesey 2002, 132). Exxon's advertorials enact a role of Exxon as a responsible corporate actor that is a protector of life, while portraying climate scientists in a negative manner (Livesey 2002, 132–141).

In contrast to the wealth of studies involving US fossil fuels corporations, research literature on climate-change-related corporate discourse by similar Australian corporations appears much more limited (Liu *et al.* 2016). The literature involves Australian corporate reports involving sustainability (Nyberg and Wright 2015), voluntary climate-change-related reporting by Australian government-owned corporations (Liu *et al.* 2016), and the construal of climate change and identity by Australian mass media (Jaspal, Nerlich, and van Vuuren 2015). Nyberg and Wright (2015) posit that voluntary reporting of carbon emissions constitutes a feature in corporate discourse in Australia. Corporate climate change discourse in Australia factors in political contestation (Nyberg and Wright 2015). Jaspal, Nerlich, and van Vuuren (2015, 4) concur with Nyberg and Wright (2015), who indicate that climate change discourse in Australia has been characterized "by engagement and disengagement, due to the perceived impact of carbon reduction for the economy, diverging political ideologies and political opportunism."

The influence of politics on corporate climate change strategies results in Australian fossil fuels corporations developing explicit climate change narratives. The narratives are organized via the frames of 'physical risk', 'regulatory risk', and 'reputational risk' (Nyberg and Wright 2015, 6). Within these frames, corporate climate change discourse is structured by means of dynamic scenarios, involving planning, developing, selling, investing, and improving measures to mitigate and offset negative consequences of climate change. Additionally, an important aspect of corporate climate change discourse in Australia involves the narrative of climate change as meaningful and profitable in a business sense and, consequently, as having monetary value (Nyber and Wright 2015, 13). By construing climate change within the frame of risk, Australian corporations make climate change meaningful and valued through calculative and monetary terms, with economic cost-benefit analyses taken into account (Nyberg and Wright 2015, 18–19).

Climate change discourse by Australian government-owned corporations is associated with the concepts of corporate image and legitimacy (Liu *et al.* 2016). In Australia, a positive corporate image in terms of environmental protection, sustainability, and climate change mitigation feeds into the corporation's legitimacy (Deegan and Rankin 1996). It is suggested that Australian corporate actors provide environmental information that is favorable to their image (Deegan and Rankin 1996, 62).

Echoing Deegan and Rankin's (1996) findings, Liu *et al.* (2016) argue that climate-change-related reports by Australian corporate actors create a positive corporate image attained via the narrative of legitimacy. Similar findings are reported by Haque, Deegan, and Inglis (2016, 24), who suggest that "Australian companies are generally more likely to disclose 'good' or 'positive' information, rather than 'bad' or 'negative' information'." Australian climate change discourse is motivated by the overarching aim of legitimation (Liu *et al.* 2016). Legitimation presupposes that a corporate actor creates a discursive context that facilitates the public's perception of the corporation's existence. However, another discursive strategy adopted by a number of Australian corporations involves environmental reporting that either deemphasizes the issue of climate change or makes no reference to it (Liu *et al.* 2016).

Previous Research Involving Conceptual Metaphors in Climate Change Discourse

Before we proceed to the role of conceptual metaphors in climate change discourse, it seems pertinent to elaborate upon the notion of conceptual metaphor. Conceptual metaphors are thought to be cognitive and social devices, seen as the result of mappings across two conceptual domains: the conceptual source domain and the conceptual target domain (Nerlich, Hamilton, and Rowe 2002, 92). In cognitive linguistics, the human conceptual system is theorized to be metaphorical in nature (Lakoff 2016, 2). Conceptual metaphors are said to permeate daily conventional language, so that all abstract reasoning is performed via metaphors. In this respect, Lakoff (1993, 244) posits that conceptual metaphors constitute "the main mechanism through which we comprehend abstract reasoning." Seen from the perspective of cognitive linguistics, conceptual metaphors are regarded as a basic structural form of experience, which organizes the engagement with and understanding of the world (Audebrand 2010, 425).

Conceptual metaphors facilitate the comprehension and communication of complex phenomena by referencing the frameworks of understanding that are mutually comprehensible (Kövecses 2013). The facilitation is enabled by projecting knowledge about the known, familiar domains of experience onto an unknown, more abstract domain. To reiterate, conceptual metaphors are regarded as a product of the transfer between the two unrelated domains. Specifically, "a process of metaphoric transfer is . . . engaged in where two separate and apparently dissimilar domains are brought into cognitive and emotional association where one is used as a framework for understanding the other" (Cornelissen and Harris 2001, 53).

By concretizing and simplifying complex abstract notions and processes, conceptual metaphors play a central role in public discourse (Charteris-Black and Musolff 2003, 153). Conceptual metaphors contribute to the understanding of complex concepts by means of transforming them into stereotypical schemas. Hence, conceptual metaphors partake in chunking the reality into meaningful and 'cognition-friendly' segments.

It follows that by providing conscious and/or subconscious anchors in the accessible and simplified form of stereotypes, conceptual metaphors can "orient institutions and individuals to particular possibilities for action and have an effect on material economic investment and policy" (Nerlich 2012, 35). Hence, institutional and corporate actors are influenced by those conceptual metaphors, which they employ in their discourse (Milne, Kearins, and Walton 2006, 810).

Conceptual metaphors not only orient individual, group, and institutional actors; they influence them to such an extent that they impose a particular aspect of the reality as well as a particular worldview (Lakoff and Johnsson 1980, 157). Arguably, conceptual metaphors reinforce a certain type of a perspective, which, in its turn, serves as a scaffolding in cognitive construals of a certain ideology. In this sense, conceptual metaphors constitute a sociocultural relation of power (Koller 2005, 201), since they can foreground some features of reality and deemphasize other features (Lakoff and Johnsson 1980, 157). In terms of discursive power relationships, metaphors frame and organize shared narratives and embed them into discursive networks of power (Koteyko, Brown, and Crawford 2008, 244).

Metaphors are commonplace in environmental and climate change discourse (Coffey 2016, 203). A cognitive-discursive account of conceptual metaphors in climate change discourse is well represented (Atanasova and Koteyko 2015; Brooks 2005; Nerlich 2010; Shaw and Nerlich 2015). Shaw and Nerlich (2015, 35) suggest that conceptual metaphors "enable as well as constrain the ways we think about policy issues, especially with regard to largely abstract, complex and seemingly intractable problems like climate change." This contention is evocative of Livesey's (2002) claim that climate change discourse by fossil fuels corporations appears to involve conceptual metaphors. Similar to fossil fuels corporations, media discourse involving climate change uses metaphors to conceptualize climate change phenomena, climate change science, and the measures aimed at climate change mitigation and adaptation (Atanasova and Koteyko 2015, 1).

Climate change criticism and skepticism on the part of US fossil fuels corporations have been mentioned in the previous section of this chapter. Elaborating upon climate change skepticism, Nerlich (2010) posits that climate change denial is metaphorically framed as 'Climategate'. This metaphor is instantiated by a mapping from the domain 'Watergate' (a 1970s political scandal in the United States) onto the denial of climate science. Thus, the mapping results in a metaphoric frame, which involves the concepts 'scandal', 'climate change', 'climate skepticism', etc. Based upon the notion of 'Climategate', Nerlich (2010) has discovered several types of conceptual metaphors associated with it. These conceptual metaphors relate to the concepts of religion, faith, church, and cults (Nerlich 2010, 11). Specifically, Nerlich has identified the conceptual metaphor 'Science as Religion', which frames the 'Climategate' narrative (Jaspal and Nerlich 2012).

Research by Atanasova and Koteyko (2015) supports Nerlich's (2010) findings involving the conceptual metaphor 'Science as Religion' in relation to the issue of climate change. Extending the framing of climate change via '(Climate) Science as Religion', previous research in the cognitive linguistic paradigm distinguishes a number of related conceptual metaphors, which bear religious connotations—e.g., 'Climate Change as a Myth' (Brooks 2005; Nerlich 2010) and 'The Earth Is a Garden'(Brooks 2005).

Russill (2015) examines debates involving climate change through the lenses of the metaphor 'Climate Change as a Tipping Point'. The tipping points are referred to as threats posed by extreme events—e.g., the melting of the Arctic ice (2015, 428). Climate change tipping points are metaphoric construals used in scientific discourse for characterizing the distinctive threats posed by abrupt change (Russill 2015, 429). The metaphor 'Climate Change as a Tipping Point' is believed to involve "both metaphor and mathematical formalization, both cultural connotations and precise technical meanings, and the semantic confusion that has accompanied their wider use" (Russill 2015, 432). The concept of threat, elaborated by Russill (2015), is present in the conceptual metaphor 'Climate Change as a Threat to the Economy' (Brooks 2005; Nerlich 2012, 33). It is emphasized that climate change discourse is dominated by the 'Market', 'Economy', and 'Threat' metaphors (Nerlich 2012). The 'Market' and 'Economy' metaphors are related to the market-based solutions to climate change mitigation and encompass the concepts of ecologically friendly economy, cost-benefit solutions in climate change mitigation (Nerlich 2012), as well as the monetary value associated with climate change adaptation (Kapranov 2015).

A Qualitative Study of Conceptual Metaphors Associated with Climate Change in Corporate Reports by ExxonMobil (United States) and BHP Billiton (Australia)

The present study regards conceptual metaphors as a framing device, which structures climate change discourse. Following Craig and Amernic (2004, 826), metaphors in discourse are considered to function as frames, which facilitate (i) the interpretation of an event or a narrative as well as (ii) the emphasis on a particular element of the discourse narrative presenting it in a positive or a negative manner. Framing involves synergies and conflicts of multiple actors under multiple realities (Keenan, King, and Willis 2016, 263). Applied to the issue of climate change, framing construes multiple realities, such as climate change mitigation, adaptation to its consequences, and/or resilience to its negative consequences (Keenan, King, and Willis 2016, 264–265).

The approach adopted in the present study subscribes to Nerlich's (2012, 35) and Boulton's (2016) views of conceptual metaphors as a deep conceptual structure—i.e., a frame. Conceptual metaphors as a frame are implicit and subconscious, and they involve the actor's guiding worldview, identity, and values (Boulton 2016, 2). The use of conceptual metaphor as a frame in a climate discourse narrative allows fossil fuels corporations to choose and manipulate aspects of the perceived reality in order to emphasize its salience to promote a particular problem definition, causal interpretation, moral evaluation, as well as a potential solution of the problem (Entman 1993, 52).

The selection of salience and its degree in a frame is dynamic, with individuals, groups, or corporate actors being able to adjust the foregrounded, salient element in the frame to suit any given discourse situation, or to provide different responses to the same climate change narrative (Fleming, Rickards, and Dowd 2015, 202). Arguably, the dynamics in the frame may lead to a number of qualitatively different conceptual metaphors associated with climate change depending on discursive preferences, contexts, and other variables involved in climate change discourse (Fleming, Rickards, and Dowd 2015). Extending this argument further, I assume that fossil fuels corporations as 'framers' structure their climate change discourse according to their specific needs.

Presumably, an individual fossil fuels corporation's unique contextual situation maps onto conceptual metaphors that are specific to the corporation's discourse. However, previous qualitative studies involving conceptual metaphors in climate change discourse indicate that international corporations—e.g., BP and Shell—frame their climate change narratives via similar and identical metaphors, such as 'Corporate Response to Climate Change as a Journey', 'Corporation as a Climate Change Concerned Citizen', and 'Corporate Response to Climate Change as Growth' (Kapranov Forthcoming). These conceptual metaphors facilitate the creation of a common discursive space between BP and Shell in terms of the issue of global climate change.

The aforementioned findings apply to the international corporations with headquarters in the Northern Hemisphere. However, little is known about potential differences or similarities in the framing of climate change discourse by the corporations representing the Southern Hemisphere—e.g., Australia. Knowledge of convergence and/or divergence of conceptual metaphors that frame climate change discourse by the fossil fuels corporations representing Australia and the United States will provide an insight into these corporations' activities related to the issue of climate change.

Hypothesis and the Specific Research Aims

Based upon previous research findings (Kapranov Forthcoming), it is assumed that corporate discourse involving climate change by Exxon and BHP will exhibit the characteristics of qualitative similarities. It is hypothesized that Exxon's and BHP's climate change discourse will be framed by qualitatively similar and/or identical conceptual metaphors. Hence, the specific research aims of the study are formulated as follows:

- To identify conceptual metaphors associated with climate change in corporate reports by Exxon and BHP;
- To juxtapose conceptual metaphors associated with climate change in corporate reports by Exxon and BHP in order to elucidate how the to-be-identified conceptual metaphors relate to the potential differences and/or similarities between the two corporations.

Materials

The materials of this qualitative study involve the official websites of Exxon and BHP (www.exxonmobil.com and www.bhpbilliton.com). In total, 14 corporate reports available at www.bhp.com and 10 corporate reports at www.exxonmobil.com have been selected for analysis. To ensure comparability and style uniformity, texts related to the topic of climate change have been garnered only from the official websites of BHP and Exxon. The texts are all considered to conform to the style of official business communication. They are all written in English and are deemed comparable for the purposes of the present study.

Procedure and Methods

An electronic search for the key phrase 'climate change' was conducted on Exxon's and BHP's official websites. The search was filtered to include the materials published within the time frame January 1, 2015–July 1, 2016.

The methodology of conceptual metaphor identification was based upon the cognitive linguistic approach, where conceptual metaphors were identified on the basis of cross-domain mappings between two unrelated domains of experience (Lakoff 1993). The procedure of conceptual metaphors identification was identical to that described in Nerlich and Jaspal (2013).

Results and Discussion

The results of the qualitative data analysis have been summarized in Table 6.1 and Table 6.2.

As is evident from the data analysis, the conceptual metaphors involved in the framing of climate change discourse by BHP and Exxon reveal a complex discursive space. BHP's and Exxon's discursive spaces associated with climate change are framed by a number of identical conceptual metaphors—for instance, 'Climate Change as a Challenge', 'Climate Change as a Risk', 'Climate Change as a Low-Carbon Future', and 'BHP/Exxon as a Climate Change Aware and Responsible Corporation'. Despite these similarities, Exxon's discursive space is characterized by the corporation-specific conceptual metaphors—e.g., 'Climate Change as an Energy-Efficient Future' and 'Climate Change as Natural Gas Future'.

Climate Change as a Challenge

Example 1 in Table 6.1 illustrates the framing of climate change by BHP as a challenge, whereby the conceptual structures of the domain 'Challenge' are mapped onto the domain 'Climate Change'. While the mapping is unidirectional—i.e., from 'Challenge' onto 'Climate Change'—it appears to involve two aspects of the domain 'Challenge', referred to in Example 1 as "these challenges." One of the challenges consists in securing access to affordable energy by the growing global population, while another challenge involves the Kyoto target of keeping the temperature increase below two degrees Celsius relative to the preindustrial level. The construals of affordable energy and the Kyoto target from the domain 'Challenge' are mapped onto the domain 'Climate Change'.

The present data support previous findings (Haque and Deegan 2010, 317) indicating that climate change and associated risks are recognized by Australian corporate actors as one of the most important business challenges. The challenge is posed not only by climate change per se but by the pressure regarding corporate climate-change-related activities from the public and government bodies.

Table 6.1 Conceptual Metaphors in Corporate Reports and Disclosures by BHP from January 1, 2015, until July 1, 2016

#	*Conceptual Metaphor*	*Example*
1.	Climate Change as a Challenge	. . . **limiting climate change** requires the global average temperature increase to remain below 2°C relative to pre-industrial levels. Successfully addressing **these challenges** will result in substantial changes in the global economy (BHP, 2015).
2.	Climate Change as a Risk	. . . we acknowledge that **climate change** will continue to have serious implications for our customers, the community and the economy. These are **risks** we need to manage (BHP, 2015).
3.	Climate Change as a Low-Carbon Future	. . . we also have a broader role to play in . . . **the development of low-emissions technology**, and sharing our market experience to support governments in delivering the changes in policy and regulation required **to successfully address climate change** (BHP, 2015).
4.	Climate Change Mitigation as a Path	We support the bipartisan objective that international negotiations in Paris deliver a positive outcome **to put the world on the path to limiting global temperatures** to less than two degrees Celsius (BHP, 2015).
5.	Climate Change as Costs	. . . achieving this goal will require an urgent and fundamental departure from business as usual, and the longer the world waits to take action, **the more it will cost** and the greater the technological, economic, social and institutional challenges we will face (BHP, 2015).
6.	Climate Change as Knowledge	BHP Billiton and Saskatchewan-based electricity provider SaskPower have signed an agreement to establish the BHP Billiton SaskPower **Carbon Capture Knowledge Centre**. . . . By enhancing **global access to the data, information and lessons learned from SaskPower's unique facility**—the first power project to successfully integrate CCS—we aim to stimulate broader deployment of the technology (BHP, 2016).
7.	BHP as a Climate Change Aware and Responsible Corporation	. . . **we recognise our responsibility** to take action by focusing on reducing our emissions, increasing our preparedness for physical climate impacts and **working with others, including industry and government, to enhance the global response to climate change**. . . . **We are committed** to continued focus on the delivery of GHG reduction opportunities within our Business (BHP, 2015).
8.	Climate Change as an Energy-Efficient Future	No direct reference
9.	Climate Change as Natural Gas Future	No direct reference

Table 6.2 Conceptual Metaphors in Corporate Reports and Disclosures by ExxonMobil from January 1, 2015, until July 1, 2016

#	*Conceptual Metaphor*	*Example*
1.	Climate Change as a Challenge	Society faces a dual energy **challenge:** We need to expand energy supplies to support economic growth and improve living standards, and we must do so in a way that is environmentally responsible. One of the most critical parts of meeting this dual **challenge** is to understand and address the risks posed by **climate change** (Exxon, 2015).
2.	Climate Change as a Risk	**The risk of climate change** is clear and **the risk** warrants action. . . . There is a broad scientific and policy consensus that action must be taken to further quantify and assess **the risks** (Exxon, 2016).
3.	Climate Change as a Low-Carbon Future	**We will also continue** to explore opportunities **to lower our greenhouse gas emissions** at every link in the energy chain across our entire business (Exxon, 2016).
4.	Climate Change Mitigation as a Path	No direct reference
5.	Climate Change as Costs	No direct reference
6.	Climate Change as Knowledge	No direct reference
7.	Exxon as a Climate Change Aware and Responsible Corporation	ExxonMobil is **a responsible participant** in the discussion on climate change—we will continue **to** research the issue, support energy efficiency, work to reduce emissions, pursue new technologies and advocate for effective policy approaches (Exxon, 2016).
8.	Climate Change as an Energy-Efficient Future	It will also require **energy to be used more efficiently** across all sectors. **Efficiency gains result from technological advances and investment.** We estimate that the **energy savings** to society associated with **efficiency gains from 2014 to 2040** will be considerable (Exxon, 2016).
9.	Climate Change as Natural Gas Future	. . . we see **natural gas** as an abundant, reliable, and clean source of energy that can meet growing power-generation needs, while **reducing greenhouse gas emissions levels** (Exxon, 2015).

Climate Change as a Risk

The framing of climate change as the 'Risk' metaphor is done by mapping from the domain 'Risk' onto the domain 'Climate Change'. It is common in corporate discourse to make references to risk and risk management. Mapped onto the domain 'Climate Change', 'Risk' foregrounds the notion of "serious implications" (BHP 2015). The seriousness of climate-change-related risks is assessed by BHP (2015) within the following two contexts: first, within the corporation's climate change awareness (e.g., "we acknowledge that climate change will continue to have serious implications for our customers, the community and the economy") and, second, within the context of corporate responsibility (as evident from the reference to "our customers, the community and the economy" [BHP 2015]). By making references to the first-person plural (e.g., 'we' and 'our' in "we acknowledge" and "our customers"), BHP positions itself as a socially situated corporation. The situatedness is entrenched into the fabrics of society, as seen from the mention of "the community." Presumably, the foregrounding of the social and societal aspects of climate-change-related risks forms BHP's strategy to portray itself as a socially responsible corporation that cares for the communities it operates in.

Framing its climate change discourse via the conceptual metaphor 'Climate Change as a Risk', BHP foregrounds a salient aspect of this framing represented by risk manageability: "These are risks we need to manage" (BHP 2015). These findings lend support to previous research by Haque, Deegan, and Inglis (2016), who suggest that Australian corporations appear to factor in the risks that climate change poses in their business practices. Climate-change-related risks manageability is incorporated into policies, and procedures are incorporated within corporate governance practices (2016).

Climate Change as a Low-Carbon Future

The conceptual metaphor 'Climate Change as a Low-Carbon Future' is instantiated by the mapping from the domain 'Low Carbon' onto the domain 'Climate Change'. The framing of the issue of climate change by means of this metaphor suggests the foregrounding of the role of low carbon usage in decreasing the corporation's carbon footprint. The decrease in GHG emissions is directly related to reducing carbon gases in the atmosphere. Thus, the development and exploitation of low-emission technology in the foreseeable future is seen by BHP as a facilitative measure to offset the negative consequences of global warming. However, it is inferred from BHP's discourse that low-emission technology is in the processes of being developed and implemented to secure a low-carbon future, as seen from Example 3 of Table 6.1: "As a leading global resources company, we also have a broader role to play in . . . the development of low-emissions technology" (BHP 2015).

The framing of climate change discourse by BHP via the conceptual metaphor 'Climate Change as a Low-Carbon Future' is evocative of a contention that climate change mitigation is metaphorically compared to a low-carbon diet (Nerlich, Evans, and Koteyko 2011). Similarly, Koteyko, Thelwall, and Nerlich (2010) have found that the metaphorical framing of climate change mitigation involves the concept of 'low carbon'. The concept is linguistically instantiated as a series of lexical compounds—e.g., 'low-carbon diet', 'carbon sinner', 'carbon detox', 'carbon future', 'carbon guilt', 'carbon morality', etc. (Koteyko, Thelwall, and Nerlich 2010, 33). Additionally, the identification of the conceptual metaphor 'Climate Change as a Low-Carbon Future' relates to Cohen (2011, 201), who indicates that climate change discourse is framed by the concept of carbon emission 'rationing'. According to Cohen (2011), this concept is evocative of the rationing of food supplies that Great Britain experienced during the Second World War (WWII). It is argued that today's limitation of carbon footprints and a cap on GHG emissions are comparable to food rationing during WWII.

Climate Change Mitigation as a Path

The mapping of conceptual structures from the domain 'Path' onto the domain 'Climate Change' gives rise to the conceptual metaphor 'Climate Change as a Path', as seen from Example 4 of Table 6.1: "to put the world on the path to limiting global temperatures to less than two degrees Celsius" (BHP 2015). Arguably, BHP frames its climate change discourse as a process, a journey to be taken on the path to a low-carbon future. The 'Path' metaphor embraces change, as opposed to the more static concepts, such as the final result (Milne, Kearins, and Walton 2006, 810). This conceptual metaphor involves a mental schema Source–Path–Goal (Result): where Source is the current situation involving climate change that needs to be mitigated and adapted to, where Path is the measures and processes of climate change mitigation (which need time to unfold), and where Goal is represented by low-carbon GHG emissions and a temperature decline (Kapranov Forthcoming). The presence of the conceptual metaphor 'Climate Change as a Path' corroborates previous findings (Kapranov 2015) that indicate that the 'Path' metaphor is identified in political discourse involving renewable energy sources within the context of climate change mitigation.

Climate Change as Costs

The conceptual metaphor 'Climate Change as Costs' is instantiated by the mapping of conceptual elements from the domain 'Costs' onto the domain 'Climate Change'. The framing can be explained by the monetary cost involved in the corporate efforts to adapt to and mitigate the negative consequences of climate change. These costs are related to investments into carbon-friendly technology and more broadly into socioeconomic changes to be undertaken in order to address the issue of climate change, as seen from Example 5 of Table 6.1: "and the longer the world waits to take action, the more it will cost and the greater the technological, economic, social and institutional challenges we will face" (BHP 2015). These findings provide support for previous research by Haque, Deegan, and Inglis (2016), who posit that climate change affects corporate profits by the increasing cost of carbon and the costs associated with physical weather and climatic impacts. In addition to the material costs, however, BHP's climate change discourse appears to foreground the notion of social costs, which, presumably, do not involve only material or monetary value.

Climate Change as Knowledge

The conceptual metaphor 'Climate Change as Knowledge' is created through the mapping from the domain 'Knowledge' onto the domain 'Climate Change'. Arguably, the framing of climate change discourse by the 'Knowledge' metaphor facilitates the construal of a corporate image that imparts the idea of corporate authority based upon science—e.g., "By enhancing global access to the data, information and lessons learned from SaskPower's unique facility—the first power project to successfully integrate CCS—we aim to stimulate broader deployment of the technology" (BHP 2016). Arguably, the presence of the conceptual metaphor 'Climate Change as Knowledge' facilitates the creation of a positive corporate image and contributes to the stakeholders' understanding of the corporate actions aimed at addressing the issue of climate change. These findings support a contention that climate change is a knowledge domain that is global in scope, effects, and importance (Czerniewicz, Goodier, and Morrell 2016, 14).

The self-image of BHP as a knowledge-based and responsible corporation corroborates previous research by Deegan and Rankin (1996, 60), who report that Australian corporate actors provide positive environmental news while being poor environmental performers. Consistent with previous findings by Deegan and Rankin (1996, 59), it is observed in the present data that BHP, as any other Australian corporation, appears to be reluctant to disclose negative environmental information.

BHP and Exxon as Climate Change Aware and Responsible Corporations

The mapping of conceptual structures from the domains 'Awareness' and 'Responsibility' onto 'Climate change' instantiates the conceptual metaphor 'BHP/Exxon as a Climate Change Aware and Responsible Corporation.' The awareness of climate change is concurrent with the recognition of each corporation's responsibility associated with fossil fuels production and consumption, as seen in Example 7 of Table 6.1: "As a major producer and consumer of fossil fuels, we recognise our responsibility to take action by focusing on reducing our emissions, increasing our preparedness for physical climate impacts and working with others, including industry and government, to enhance the global response to climate change" (BHP 2015).

As the Australian economy becomes more resource dependent (Cleary 2016, 152), the role of the fossil fuels corporations—for instance, BHP—in GHG emission and in industrial pollution grows exponentially. Consequently, it is in the corporation's interest to dissociate itself from the negative role of industrial polluter. One of the measures in such a dissociation process is to publicly recognize the responsibility for GHG emissions. Concurrent with exhibiting public awareness of the negative consequences of anthropogenic climate change, climate change discourse by BHP is suggestive of presenting itself in a positive light. This measure involves the corporation's further commitment to climate change mitigation—e.g., "We are committed to continued focus on the delivery of GHG reduction opportunities within our business" (BHP 2015). The present findings seem to support a contention that climate change is a critical issue for fossil fuels corporations since their GHG emissions contribute to anthropogenic climate change, thus requiring solutions to mitigate it (Nyberg and Wright 2015).

Arguably, the strategy of climate change awareness taken in conjunction with climate change responsibility construe BHP as a socially responsible corporation, caring for the environment. If we extend this suggestion further, it can be assumed that the corporation acts as a responsible 'Citizen' who is climate change aware. Corporate citizenship is commonly used by major international corporations (Moon, Crane, and Matten 2005, 429). In relation to the issue of climate change, the corporate 'Citizen' strives to preserve the environment and to offset negative consequences of climate change (Audebrand 2010).

Previous studies indicate that Exxon's corporate discourse is marked by climate change awareness and responsibility, whereby the corporation frames its understanding of climate change as a complex issue associated with significant risks for humans and ecosystems (Lorenz, Rabinsky, and Wurtzel 2008, 2). The present qualitative data analysis supports research by Lorenz, Rabinsky, and Wurtzel (2008)—e.g., "ExxonMobil is a responsible participant in the discussion on climate change" (Exxon 2016). Presumably, the conceptual metaphors 'Climate Change as Knowledge' and 'BHP/Exxon as a Climate Change Aware and Responsible Corporation' contribute to the legitimation of the corporation in the sense of Liu *et al.* (2016).

Climate Change as an Energy-Efficient Future

The mapping of conceptual elements from the domain 'Energy Efficiency' onto 'Climate Change' results in conceptual metaphor 'Climate Change as an Energy-Efficient Future'. This metaphor is Exxon-specific—i.e., it is absent in the corpus of BHP's reports. Exxon's focus on energy efficiency as a means to reduce GHG emissions and as a measure of climate change mitigation in the future is seen in Example 8 of Table 6.2: "It will also require energy to be used more efficiently across all sectors. Efficiency gains result from technological advances and investment. We estimate that the energy savings to society associated with efficiency gains from 2014 to 2040 will be considerable" (Exxon 2016).

The concept of energy efficiency as a salient feature of Exxon's climate change discourse is reminiscent of the contention that Exxon has traditionally been reliant on safeguarding the status quo in its corporate strategies (Livesey 2002). It is inferred from the present data that the focus on an energy-efficient future presupposes reliance on fossil fuels, which will be used more efficiently compared with today's usage. In other words, the status quo involving fossil fuels will be preserved by Exxon in the future. This line of argument has not been identified in the discursive space by BHP.

Climate Change as Natural Gas Future

The conceptual metaphor 'Climate Change as Natural Gas Future' is instantiated by the mapping from the domain of 'Natural Gas' onto the domain 'Climate Change'. Arguably, the presence of this conceptual metaphor in Exxon's discursive space symbolizes the difference between Exxon and BHP in terms of climate change mitigation. Whilst BHP has access to one of the most significant natural gas supplies in the Gascoyne region in Western Australia, the corporation, nevertheless, does not frame its climate change discourse via natural gas metaphors. BHP seems to deemphasize the role of Gascoyne natural gas as an alternative to GHG emissions. Exxon, in contrast to BHP, appears to favor the so-called 'natural gas future' as a viable measure of climate change mitigation. Exxon's approach is exemplified by Example 9 of Table 6.2: "we see natural gas as an abundant, reliable, and clean source of energy that can meet growing power-generation needs, while reducing greenhouse gas emissions levels" (Exxon 2015).

Conclusions

This chapter involves a qualitative study aimed at identifying conceptual metaphors associated with the issue of climate change in corporate reports by ExxonMobil (United States) and BHP Billiton (Australia). The focus of this study is on establishing whether or not there are qualitative differences and/or similarities in conceptual metaphors associated with climate change identified in corporate reports by these two corporations. For that purpose, corporate reports by these two corporations have been collected at the websites www.bhpbilliton.com and www.exxonmobil.com.

The data analysis reveals that the conceptual metaphors involved in the framing of climate change discourse by BHP and Exxon involve a multilayered discursive space, represented by a number of identical conceptual metaphors—e.g., 'Climate Change as a Challenge', 'Climate Change as a Risk', 'Climate Change as a Low-Carbon Future', and 'BHP/Exxon as a Climate Change Aware and Responsible Corporation'. Despite these similarities, Exxon's discursive space is characterized by the corporation-specific conceptual metaphors—e.g., 'Climate Change as an Energy-Efficient Future' and 'Climate Change as Natural Gas Future'. The differences in framing climate change discourse are suggestive of two distinct approaches adopted by these corporations. These approaches can be summarized as follows: While

ExxonMobil construes its climate change discourse via conceptual metaphors associated with natural gas and efficient energy consumption, BHP frames its climate change discourse by means of conceptual metaphors evocative of climate change mitigation as a process, a path, which requires knowledge and investment.

References

Allen, M. 2016. *Strategic communication for sustainable organizations*. Heidelberg: Springer.

Atanasova, D., and N. Koteyko. 2015. "Metaphors in *Guardian* online and *Mail* online opinion-page content on climate change: War, religion, and politics." *Environmental Communication*: 1–18. doi: dx.doi.org/10.1080/17524032.2015.1024705

Audebrand, L. K. 2010. "Sustainability in strategic management education: The quest for new root metaphors." *Academy of Management Learning and Education* 9 (3): 413–428.

BHP. 2015. Climate change: Portfolio analysis- BHP Billiton. June 1, 2016. http://www.bhpbilliton.com/society/climate-change

BHP. 2016. BHP Billiton and SaskPower establish carbon capture and storage knowledge centre. June 1, 2016. http://www.bhpbilliton.com/investors/news/bhp-billiton-and-saskpower-establish-carbon-capture-and-storage-knowledge-centre

Boulton, E. 2016. "Climate change as a 'hyperobject': A critical review of Timothy Morton's reframing narrative." *Wiley Periodicals: Climate Change* 7 (5): 772–785. doi: 10.1002/wcc.410

Brooks, C. 2005. "Who will cry for the ice? A preliminary sketch of metaphorical framing and conceptual understanding in climate change terminology." *Breslauer Symposium*. https://escholarship.org/uc/item/1v09b87p. Accessed 1 June 2016.

Charteris-Black, J., and A. Musolff. 2003. "'Battered hero' or 'innocent victim'? A comparative study of metaphors for euro trading in British and German financial reporting." *English for Specific Purposes* 22: 153–176.

Cleary, P. 2016. "Poles apart: Comparative resource sector governance in Australia and Norway." *Australian Journal of Political Science* 51 (1): 150–162.

Coffey, B. 2016. "Unpacking the politics of natural capital and economic metaphors in environmental policy discourse." *Environmental Politics* 25 (2): 203–222.

Cohen, M. J. 2011. "Is the UK preparing for 'war'? Military metaphors, personal carbon allowances, and consumption rationing in historical perspective." *Climatic Change* 104: 199–222.

Cornelissen, J., and P. Harris. 2001. "The corporate identity metaphor: Perspectives, problems and prospects." *Journal of Marketing Management* 17 (1–2): 49–71.

Craig, R. J., and J. H. Amernic. 2004. "Enron discourse: The rhetoric of a resilient capitalism." *Critical Perspectives on Accounting* 15 (6): 813–852.

Czerniewicz, L., S. Goodier, and R. Morrell. 2016. "Southern knowledge online? Climate change research discoverability and communication practices." *Information, Communication & Society* 20 (3): 386–405. doi: dx.doi.org/10.1080/1369118X.2016.1168473

Deengan, C., and M. Rankin. 1996. "Do Australian companies report environmental news objectively? An analysis of environmental disclosures by forms prosecuted successfully by the environmental protection authority." *Accounting, Auditing & Accountability Journal* 9 (2): 50–67.

Entman, R. M. 1993. "Framing: Toward clarification of a fractured paradigm." *Journal of Communication* 43 (4): 51–58.

Exxonmobil. 2015. Engaging on climate change. May 1, 2016. http://corporate.exxonmobil.com/en/current-issues/climate-policy/climate-perspectives/engagement-to-address-climate-change

Exxonmobil. 2016. ExxonMobil's perspectives on climate change. May 1, 2016. http://corporate.exxonmobil.com/en/current-issues/climate-policy/climate-perspectives

Fleming, A., L. Rickards, and A.-M. Dowd. 2015. "Understanding convergence and divergence in the framing of climate change responses: An analysis of two wine companies." *Environmental Science and Policy* 51: 202–214.

Fløttum, K., and T. Dahl. 2012. "Different contexts, different 'stories'? A linguistic comparison of two development reports on climate change." *Language and Communication* 32: 14–23.

Good, J. E. 2008. "The framing of climate change in Canadian, American, and international newspapers: A media propaganda model analysis." *Canadian Journal of Communication* 33: 233–255.

Greenberg, J., G. Knight, and E. Westersund. 2011. "Spinning climate change: Corporate and NGO public relations strategies in Canada and the United States." *The International Communication Gazette* 73 (1–2): 65–82.

Haque, S., and C. Deegan. 2010. "Corporate climate change-related governance practices and related disclosures: Evidence from Australia." *Australian Accounting Review* 20 (4): 317–333.

Haque, S., C. Deegan, and R. Inglis. 2016. "Demand for, and impediments to, the disclosure of information about climate change-related corporate governance practices." *Accounting and Business Research* 46 (6): 620–664. doi: dx.doi.org/10.1080/00014788.2015.1133276

Jaspal, R., and B. Nerlich. 2012. "When climate science became climate politics: British media representations of climate change in 1988." *Public Understanding of Science* 23 (2): 122–141. doi: 0963662512440219

Jaspal, R., B. Nerlich, and K. van Vuuren. 2015. "Embracing and resisting climate identities in the Australian press: Sceptics, scientists and politics." *Public Understanding of Science* 7: 807–824. doi: 10.1177/0963662515584287

Jones, C., and D. L. Levy. 2007. "North American business strategies towards climate change." *European Management Journal* 25 (6): 428–440.

Kapranov, O. 2015. "Conceptual metaphors in Ukrainian Prime Ministers' discourse involving renewables." *Topics in Linguistics* 16 (1): 4–16.

Kapranov, O. Forthcoming. "Do international corporations speak in one voice on climate change." To appear in *SSFH 2015* edited by K. Papaja and C. Can.

Keenan, J., D. A. King, and D. Willis. 2016. "Understanding conceptual climate change meanings and preferences of multi-actor professional leadership in New York." *Journal of Environmental Policy & Planning* 18 (3): 261–285.

Koller, V. 2005. "Critical discourse analysis and social cognition: Evidence from business media discourse." *Discourse and Society* 16 (2): 199–224.

Koteyko, N., B. Brown, and P. Crawford. 2008. "The dead parrot and the dying swan: The role of metaphor scenarios in UK press coverage of Avian Flu in the UK in 2005–2006." *Metaphor and Symbol* 23 (4): 242–261.

Koteyko, N., M. Thelwall, and B. Nerlich. 2010. "From carbon markets to carbon morality: Creative compounds as framing devices in online discourses on climate change mitigation." *Science Communication* 32 (1): 25–54.

Kövecses, Z. 2013. "The metaphor-metonymy relationship: Correlation metaphors are based on metonymy." *Metaphor and Symbol* 28: 75–88.

Lakoff, G. 1993. "The contemporary theory of metaphor." *Metaphor and Thought* 2: 202–251.

Lakoff, G. 2016. "Language and emotion." *Emotion Review* 8 (3): 269–273. doi: 10.1177/1754073915595097

Lakoff, G., and M. Johnsson. 1980. *Metaphors we live by*. Chicago: University of Chicago Press.

Liu, Z., S. Abhayawansa, C. Jubb, and L. Perera. 2016. "Voluntary climate change-related reporting of Australian government-owned corporations: The impact of the National Greenhouse and Energy Reporting Act 2007." *Proceedings of 39th Annual Congress of the European Accounting Association*, Maastricht.

Livesey, S. 2002. "Global warming wars: Rhetorical and discourse analytic approaches to ExxonMobil's corporate public discourse." *The Journal of Business Communication* 39 (1): 117–148.

Lorenz, S., M. Rabinsky, and J. Wurtzel. 2008. "Exxon mobile and British petroleum: A tale of two companies or just business as usual?" *Natural Resources and Environment* 501: 1–21.

Milne, M., K. Kearins, and S. Walton. 2006. "Creating adventure in Wonderland: The journey metaphor and environmental sustainability." *Organization* 13 (6): 801–839.

Moon, J., A. Crane, and D. Matten. 2005. "Can corporations be citizens? Corporate citizenship as a metaphor for business participation in society." *Business Ethics Quarterly* 15 (3): 429–453.

Nerlich, B. 2010. "'Climategate': Paradoxical metaphors and political paralysis." *Environmental Values* 19 (4): 419–442.

Nerlich, B. 2012. "'Low carbon' metals, markets and metaphors: The creation of economic expectations about climate change mitigation." *Climatic Change* 110: 31–51.

Nerlich, B., V. Evans, and N. Koteyko. 2011. "Low carbon diet: Reducing the complexities of climate change to human scale." *Language and Cognition* 3 (1): 45–82.

Nerlich, B., C. A. Hamilton, and V. Rowe. 2002. "Conceptualising foot and mouth disease: The socio-cultural role of metaphors, frames and narratives." *Metaphorik.de* 2: 90–108.

Nerlich, B., and R. Jaspal 2013. "UK media representations of carbon capture and storage: Actors, frames and metaphors." *Metaphor and the Social World* 3 (1): 35–53.

Nyberg, D., A. Spicer, and C. Wright. 2013. "Incorporating citizens: Corporate political engagement with climate change in Australia." *Organization* 20(3): 433–453.

Nyberg, D., and C. Wright. 2015. "Performative and political: Corporate constructions of climate change risk." *Organization* 23 (5): 617–638. doi: 1350508415572038

Pulver, S. 2007. "Making sense of corporate environmentalism: An environmental contestation approach to analyzing the causes and consequences of the climate change policy split in the oil industry." *Organization & Environment* 20 (1): 44–83.

Russill, C. 2015. "Climate change tipping points: Origins, precursors, and debates." *Climate Change* 6: 427–434.

Shaw, C., and B. Nerlich. 2015. "Metaphor as a mechanism of global climate change governance: A study of international policies, 1992–2012." *Ecological Economics* 109: 34–40.

Talbot, D., and O. Boiral. 2014. "Strategies for climate change and impression management: A case study among Canada's large industrial emitters." *Journal of Business Ethics* 132: 329–346.

Sources

www.bhpbilliton.com
www.exxonmobil.com

Language of Climate Action

7 Willingness of Action

Kjersti Fløttum

Introduction

Tackling climate change and transitioning to a low-carbon society require both political and individual action. This transition depends on increased alignment between decision makers' mitigation strategies (to reduce or prevent emission of greenhouse gases) and citizens' opinions about and attitudes to mitigation. However, opinions and attitudes must be accompanied by action. The need of action is now formally approved by 195 countries having signed the Paris Agreement of December 12, 2015, aiming to "strengthen the global response to the threat of climate change" (see Chapters 3 and 8 in this volume). The success of the Paris Agreement is dependent on every country's implementation of the obligations stated in the document, and such implementation rests to a large extent on the citizens' willingness to act. Mobilizing action can be made more optimal with knowledge about how the public views climate change solutions and what they think needs to be done in the face of climate change. Thus, the main purpose of this chapter is to investigate what Norwegian citizens think should be done to get to grips with climate change. Unsurprisingly, the public debate on the complex phenomenon of climate change is characterized by multiple voices expressing divergent points of view according to interests, values, and worldviews (Hulme 2009; Moser 2010; Nerlich, Koteyko, and Brown 2010). From being to a large extent focused on the causes of climate change (the attribution issue), the debate now is much more oriented toward what measures to undertake to meet the many predicted challenges at different levels (local, national, and global).

In order to investigate the issue of willingness, we have asked the following open-ended question in two representative surveys undertaken in spring and autumn 2015 by the Norwegian Citizen Panel (see next section): "Concerning climate change, what do you think should be done?" (In Norwegian: "Når det gjelder klimaendringer, hva mener du bør gjøres?) Results of these surveys are merged and will be presented and discussed in the following sections as results from one survey. The survey was conducted in Norwegian; all examples taken from the answers are translated to English in this presentation.

The literature on public opinions about climate solutions is increasing, including studies concerning to what extent one should adapt to and/or mitigate the threats (see, for example, Gellein, Kipperberg, and Risa 2015). But we do not know enough about how people would choose between adaptation and mitigation. There is also a lack of knowledge on more detailed policy preferences. However, some questions related to support for or opposition to various measures—for example, carbon taxes and subsidies of renewable energy—have been studied (see Aldy, Kotchen, and Leiserowitz 2012; Kotchen, Boyle, and Leiserowitz 2013; Krosnick and MacInnis 2013; Smith and Leiserowitz 2014; Tvinnereim and Ivarsflaten 2016; Tvinnereim and Steinshamn 2016). In addition to the question about what should or could be done, there is the issue of responsibility: Who should do what? For work on opinions about government responsibility in Europe and the United States, see the review undertaken by Lorenzoni and Pidgeon (2006). In a more recent study, Obradovich and Guenther (2016) investigate the relation between individual and collective responsibility and find that emphasizing collective responsibility for the causes of climate change increases mitigation behavior. Since many of these studies use closed-ended questions—i.e., with answer alternatives that are selected and defined by the researchers responsible for the surveys—more knowledge is needed about citizens' views formulated in their own words.

The present study is based on one open-ended question, which invites respondents to freely formulate their answers, both in form and in length. This is a choice justified by the fact that when respondents can express their views in their own words, their answers provide richer and more nuanced data than with closed-ended questions, adding great value to knowledge about people's perception of climate solutions (for studies using open-ended survey questions to investigate attitudes to climate change, see Fløttum, Dahl, and Rivenes 2016; Leiserowitz 2005; Lorenzoni *et al.* 2006; Smith and Leiserowitz 2014; Tvinnereim and Fløttum 2015). In addition, in their freely formulated answers, respondents will express what is important to them; thus, we get access to what they prioritize.

The overarching research question in the present study is what citizens think should be done about climate change and by whom; their answers will indicate to what extent they are willing to act. Our working hypothesis is that they are clear about the fact that something should be done but have divergent views on whose responsibilities the proposed actions are.

In the rest of this chapter, I will first describe, in the "Material and Method" section, the data on which the various analyses are based, as well as the methods employed. The two subsequent sections will present the analysis results. The first of these sections reports on the citizens' topic prevalence concerning climate solutions, while the second presents findings from a linguistic focus on questions related to obligation (deontic perspective), to responsibilities (actor perspective), and to the deliberation between different points of view expressed in the respondents' answers (polyphonic perspective). The final section of the chapter will sum up and offer some concluding remarks.

Material and Method

The material of the present work stems from data obtained through surveys (spring and autumn 2015; hereafter referred to as one survey) undertaken by the probability-based representative internet panel—the Norwegian Citizen Panel (NCP), part of the Digital Social Science Core Facility (DIGSSCORE) established at the University of Bergen (see http://digsscore.uib.no/ and http://digsscore.uib.no/panel). The participants were asked this open-ended question: "Concerning climate change, what do you think should be done?" It yielded answers. There was great variation in the answers concerning both length and form or construction. The length varies from 1 to approximately 146 words (median length: 14). The way they constructed their answers can be divided into three main categories:

(1) REJECTION: Respondents express a non-acceptance of the question by a simple "Nothing" or similar expressions, or just "Don't know."
(2) LIST: Respondents enumerate several measures/instruments (such as "Better public transport. Cheaper tickets. Car free days/car free zones") without further comments (some answers consisted of only one measure—e.g., "Reduce emissions").
(3) STORY: Respondents construct long answers constituting complete stories or narratives, including discussion related to both complicating issues (typically climate change and non-preferred solutions) and preferred solution(s). Here is an example:

> "Follow up paragraph 112 of the Constitution in practice, i.e. move the burden of proof so that sustainable solutions are mandatory unless there are special reasons not to choose them, not like now, where you choose traditionally (and what you think is the cheapest solution in the short term), instead of robust long-term solutions—which are also required to develop the market and industry so that they are capable of coming up with real and good sustainable solutions. There is also a need for policies which make us not go for "low-hanging fruit" (i.e. fairly simple solutions), which block the way to solutions which might yield energy savings of up to 5–8%. I think old men (at least some of them) must be ejected from the decision-making process."

In the first part of the analyses, we make use of a semi-automated quantitative text analysis technique named Structural Topic Modelling (STM) (Roberts *et al.* 2014) in order to systematize the content of the large and very heterogeneous text material that resulted from the survey. This recent development in machine learning based on analysis of textual data allows for an inductive search for distinct topics in the text material; in other words, we discover topics instead of assuming them (Roberts *et al.* 2014, 1066). STM generates topics using a clustering algorithm based on the frequency and co-occurrence of words across documents. The result is various topic prevalences, defined as the degree to which a single response belongs to a given topic, based on the words it contains. The topic induction is followed by thorough qualitative examination of a large set of possible topic configurations. In this case, we ended up with a set of seven topics broadly covering the answers given by the respondents. I will come back to these in the "Analyses" section. For more details about the semi-automated analysis, the subsequent qualitative analysis, and the possibilities to correlate topic prevalence with individual-level background variables such as demographics and political attitudes, see Tvinnereim and Fløttum (2015) and Tvinnereim *et al.* (Forthcoming).

The second part of the analyses consists of various linguistic approaches, mainly with a point of departure in the frequency and concordances of a selection of lexical-semantic elements. In order to identify and characterize salient patterns of the selected words, we generated concordances—i.e., the close textual surroundings of a word, how this word pairs or co-occurs with other words. We sorted alphabetically to the right of the target word, with ten words of co-text to the left and to the right (on concordances and automated methods, see Anthony 2011; Fløttum *et al.* 2014). The questions investigated, aiming to answer the main issue of willingness of action, are related to the perspectives of obligation (what should or must be done), of responsibilities (who will/should undertake the various proposed actions or solutions), and of deliberation (to what extent do divergent views appear). The main theoretical inspiration is the theory of linguistic polyphony, studying how linguistic devices indicate multivoicedness—i.e., the presence of different points of view in the speaker's own voice as well as the presence of different voices in addition to the speaker's voice (see Chapter 3 in this volume; Fløttum 2010; Fløttum and Dahl 2011, 2014; Fløttum and Gjerstad 2013; Nølke, Fløttum, and Norén 2004). The choice of this theoretical framework is based on the presence of the multiple points of view and voices in the climate change debate.

Analyses of the 'Survey Discourse'

Topic Prevalence in Citizens' Views on Climate Change Solutions

After a rigorous qualitative assessment, by five researchers, of the results provided by multiple STM machine runs, a set of seven topics was chosen for our analysis. We selected the run that was assessed to be the best corresponding to the diversity of topics identified through the manual readings.

The seven topics are the following, labeled collectively by the five researchers, with percentage of importance in parentheses and followed by the most representative answer for each topic (provided by the STM run):

(1) Developing public transport (18.5%)

"Improve public transport, i.e. lower the prices and build public transport (more tramways, buses, increase frequency of departures, etc.)."

(2) Energy transition (16.4%)

"Stop the extraction of oil, invest in research/development of environmentally friendly technology, transition to renewable energy sources, etc."

(3) Attribution of climate change (15.8%)

"Climatic changes have happened on earth throughout the ages. Warm times and cold times come and go. Experts are preaching about one threat after another, but nothing is completely certain. They have been wrong before and are not 'all-knowing'. Just remember the advice on how excellent diesel cars were, and which all politicians lauded up to the skies. What is the next thing that experts/scientists are coming up with which turns out not to be true? We all need to accept periods of change on the globe, whether we want to or not. Man predicts but God/the Universe rules."

(4) Emission reduction (14.1%)

"Reduction of emissions in the transport sector. Cleaning up emissions in the air from production. Tax 'dangerous' emissions."

(5) International collaboration and responsibilities (13.1%)

"Norway is so small, the big environmental villains like China and other countries should implement measures at once. Norway can contribute by spreading knowledge and information."

(6) Changes in consumption and lifestyle (12.6%)

"Less driving, fewer long flights vacation, grow more organic food."

(7) Governmental measures/instruments (9.6%)

"It is not possible to make actors choose climate friendly alternatives unless it also makes sense economically. That is why it is necessary to implement political tools which make climate friendly solutions economically friendly solutions. The environmental costs should be placed with those who cause them, i.e. taxes etc."

Although these examples give only a very limited impression of the diversity in both content and form of the answers, the list of topics provides an overview of the respondents' priorities of solutions to climate change. We observe that they are fairly well distributed, with a range from 18.5% to 9.6%. 'Public transport' is the 'winner'—i.e., the topic that is most frequently prioritized among the respondents. This corresponds well to questions that are often raised in media and the public debate. 'Public transport' and 'Change in consumption and lifestyle' are perhaps the most concrete topics and the closest to people's needs and daily activities. 'Change in consumption and lifestyle' is the topic that conveys in the clearest way people's readiness or willingness to act. At the other end, we find 'Energy transition', 'Governmental measures/instruments', and 'International collaboration and responsibilities', representing more complex issues that are not directly related in the same way to individuals' daily needs and concerns. The topic 'Emission reduction' is found roughly midrange, a concern that may be seen as the responsibility of individuals, corporations, and national governments. We also note the important place of the topic 'Attribution of climate change', often expressing a rejection of the question, which in fact presupposes that something should be done, or a discussion of the causes of climate change, questioning human influence. In general, through all the seven topics, we see that Norwegians give special importance to mitigation; action related to adaptation to climate change is almost absent, even though Norway has experienced several dramatic extreme weather events that may be related to climate change. (For further discussion of these topics, see Tvinnereim *et al.* Forthcoming.)

Within the answer examples of the seven topics presented above, a range of linguistic forms are observed. There are imperative or infinitive verb forms, such as *improve*, *stop*, *build*, and *invest*, conveying a meaning of command or obligation, also expressed through modal auxiliaries such as *should* and the evaluation *it is necessary.* We find descriptions, more or less value-laden ("Norway is so small"), and explanations ("that is why").

The next part of this analysis section will be devoted to linguistic analyses of respondents' answers with a focus on questions related to obligation (deontic perspective), to responsibilities (actor perspective), and to deliberation (polyphonic perspective).

Linguistic Analyses: Obligation, Responsibilities, and Deliberation

Obligation

The deontic perspective of obligation is already present through the modal auxiliary *should* (Norwegian: *bør*) in this open-ended question: "Concerning climate change, what do you think should be done?" For the analysis of the presence and the forms of deontic perspectives in the respondents' answers, multiple linguistic phenomena are relevant. However, we have chosen to focus on a selection of modal auxiliaries that are clear carriers of deontic meaning. They indicate how the world ought to be according to certain norms, expectations, and desires—typically expressing some obligation or directive in the form of commands and requests. When an obligation is expressed, as in "emissions should be reduced," the feasibility of the suggested actions is presupposed through 'should' (Fløttum and Gjerstad 2013; Kerbrat-Orecchioni 2002). In polyphonic terms, there is a collective voice according to which there are actors who are capable of accomplishing the task in question. There is thus not only an imposition of obligation (and responsibility) but a construction of a common point of view regarding the competence of potential actors (individual or public actors and institutions). An important question is then who these actors are and to what extent they are mentioned explicitly. I will come back to this below, under the heading "Responsibilities."

The analyses undertaken have focused on four forms of Norwegian modal auxiliaries: *bør*, *burde*, *skal*, and *må*. Modal auxiliaries are in themselves polysemantic, being able to express many different meanings. In the following, I will give a short and simplified description of what the relevant meanings for the present analysis are and will, to the extent possible, refer to English quasi-corresponding auxiliaries. For more in-depth explanations, see Faarlund, Lie, and Vannebo (1997), Lyons (1977), and Quirk *et al.* (1972).

The deontic variant of *bør/burde* expresses various forms of necessity, that somebody has the duty to do something, or that something has to be undertaken. However, the meaning is often more a recommendation or a piece of advice than an injunction. Implicitly, in the recommendation, there may be a moral or other reasons for doing what is expressed. The form *bør* is stronger than *burde* (the preterit form), which expresses a weaker directive, in a more tentative way than *bør.* These two auxiliaries correspond generally to English *should.*

The auxiliary *skal* in its deontic use expresses an injunction or a command. The closest corresponding auxiliary in English may be *shall*, in its meaning of strong volition (however, it must be noted that *shall* also can be used in a quasi-legal sense; see Quirk *et al.* 1972: 99; Chapter 8 in this volume). Norwegian *skal* is close to *må*, which denotes a clear obligation or compulsion about something one is obliged to do. This *må* may correspond to English *must.*

Even though many possible meaning nuances may have been overlooked in this simplified presentation of the auxiliaries in question, we see that they can be placed on a graded scale of obligation, from the "mildest" *burde*, via *bør* and *skal* to *må.*

The distribution of the four studied auxiliaries in the respondents' answers is as follows: *må* (1,076), *bør* (591), *skal* (185), *burde* (100). This shows that a majority (55%) of the used auxiliaries (total of 1,952) correspond to the strongest one, *må* (*must*), which indicates a clear view of obligation and necessity, and perhaps urgency.

In the following, I will focus on *bør* (*should*), which is the same auxiliary as the one used in the survey question. In order to understand how *bør* is used, it is necessary to look at its immediate surroundings, in particular the verbs and verb expressions to which the meaning of obligation is related—i.e., what actions the citizens prioritize, and what they see as requested or recommended. The concordances with *bør* as the target word show that a large variety of expressions have been used. The following verbs are the most frequent:

> *ta* (*take*): 36; *gjøre* (*do*): 34; *satse* (*go for*): 31; *redusere* (*reduce*): 25; *bli* (*become*): 25; *bidra* (*contribute*): 18; *bruke* (*use*): 16; *forby* (*prohibit*): 9; *få* (*get*): 9; *begrense* (*limit*): 8; *hjelpe* (*help*): 8; *sette* (*put*); 8; *innføre* (*introduce*): 6; *kutte* (*cut*): 6; *skje* (*happen*): 5; *prioritere* (*prioritize*): 4.

These verbs are quite clearly action oriented. Some of them convey meanings that are understandable without more context—such as *go for*, *reduce*, *prohibit*, *limit*, *help*, *cut*, and *prioritize*. However, to get the whole picture, it is necessary to look at their complements. For *go for*, it mainly concerns renewable energy and public transport; *reduce* is often linked to emissions and traffic (in particular air traffic); *prohibit*, to the use of fossil fuel; *limit*, to various factors contributing to emissions; *help*, to other countries (big polluters); *cut*, to emissions; and *prioritize*, to the reduction or cleaning of greenhouse gases.

For the other verbs, it is necessary to look further into what constructions they are part of. Concerning *take*, it is in a large number of occurrences related to "take responsibility" or "take climate changes seriously"; as regards *do*, its related expressions cover "to do what we can," from local to global levels (however, here we also find rejections such as "have no opinion of what should be done"); as for *become*, it is often related to adjectival forms with a positive meaning, as in "transport should become better" or "[people should] become more conscious" (for example, related to our habits of consumption); *contribute* is in several cases similar to *do*, in that it is followed by "with what one can / with the little one can"; *use* is followed by a variety of complements (for example, spend money on "research on green technology"); the same is the case for *get*, where complements include both "get electric driven boats and buses" and get "more balanced information"; for *put*, the constructions mainly concern putting money and/or efforts into various measures; *introduce* is accompanied by expressions related to restrictions and taxes; and, finally, *happen* co-occurs with expressions related to transport measures.

To sum up, we observe that a large heterogeneity of actions, measures, and instruments are suggested and recommended by the respondents in order to meet the many predicted challenges of climate change. The topics uncovered by the semi-automated analysis described above are clearly manifested in the answers freely formulated by the respondents.

Through the above analysis of the most frequent verb expressions related to the modal auxiliary *bør*, and their complements, the first part of the overarching research question of this study, presented in the "Introduction," can now be answered: The representative sample of Norwegian citizens included in the survey have quite clear-cut views on what should be done about climate change. However, the second part of the question, by whom the proposed measures should be undertaken, is still unanswered. This will be examined in the next part of this section.

Responsibilities

The first issue to be examined here is to what extent there are actors that are explicitly "appointed" in the respondents' answers. Who are or should be responsible for the recommended or requested actions, according to the respondents? And to what extent do the respondents consider themselves as actors and thereby manifest willingness to act?

Through the answers containing the modal auxiliary *bør* (*should*), we have an interesting indication as to what extent there are explicitly mentioned actors related to the proposed actions. Of the 248 most frequent constructions in which *bør* appears (see above), 96 are formulated in a passive form; and, of these, only 3 contain an explicitly mentioned agent. One of the verbs appearing in a passive sentence is *redusere* (*reduce*); 19 of the 25 occurrences appear in the Norwegian s-passive (reflexive) form: *reduseres* (English: *be reduced*). Here are some examples: "Air traffic/factory emissions/greenhouse gas emissions/meat production/energy consumption . . . should be reduced." It is obvious that such measures necessitate various kinds of actors in order to be undertaken, from authorities via industries to individuals. Generally, this is not expressed. One reason may be that it is difficult in many cases to decide who are or should be responsible for undertaking the proposed actions and that a mix of different collaborating actors may be required. However, in some examples, the respondents place a responsibility on (more or less) specific actors, such as in "Norway/we/people should reduce . . ."

In order to get a broader and more general picture of the presence of explicit actors in the answers, a frequency analysis of the following pronouns and nouns was undertaken (number of occurrences is in parentheses): *jeg/eg* (*I*) (310); *vi* (*we*) (981); *alle* (*all*) (442); *staten* (*the state*) (73); *myndigheter/-ene* ([*the*] *authorities*) (37); *politikere/-erne* ([*the*] *politicians*) (74).

The first-person singular pronoun *jeg* is in the majority of its occurrences related to opinion verbs such as *mene* (*think*), *synes* (*think*, *find*), and *tro* (*believe*), typically introducing a subordinate complement clause with or without *at* (*that*):

> "In Norway I think more efforts should be put into public transport."

This *I think* is just an introduction that does not bring us further in the actor question. We also find the rejection type as in:

> "I think it has to do with natural variation."

As well as a kind of defense of Norway:

> "I think Norway already does a lot . . ."

The respondents bring their voices directly into the answer when using *I*, being responsible only for their own views. With the first-person plural pronoun *vi* (*we*), this is different. When respondents use *vi*, they manifest a kind of collective and undefined voice. In the answers, the use of *vi* seems to be of the inclusive kind, but it is generally difficult to determine who are included. It could be *vi* (the people on this planet), *vi* (the Norwegians), *vi* (the non-experts), etc. In contrast to the answers with *I*, the answers with *vi* do not contain opinion verbs but contain more action-oriented verbs. In these cases, there is a dominance of modal auxiliaries with a deontic meaning being directly related to *vi*. We find all variants of the auxiliaries: *bør*, *burde*, *skal*, *må*. However, interestingly, we also find constructions with *kan* (*can*), here mostly in the sense of 'ability', as in:

> "All of us should think a bit more about what we can do in terms of 'small things' in our everyday life which can help the environment."

But we also find points of view expressing helplessness or resignation, and passivity:

> "There is little we can do . . ."

The pronoun *alle* (*all*) also manifests a collective voice and contributes to an emphasis on collective responsibility in one way or another. In order to pursue the actor perspective, *all* on its own and *all* in a noun phrase (such as 'all businesses', 'all inhabitants') in the function of syntactic subject were examined (208 occurrences). A striking characteristic of the sentences in which they appear is the co-occurrence of deontic auxiliaries. There are 43 occurrences of *alle må* (*all must*) and 27 of *alle bør* (*all should*). Typically, these are combined with the verb *contribute*, such as in:

> "All must/should contribute."

This happens sometimes without any further explanation of what to contribute with, and sometimes followed by the vague expression "with what they can do." Another typical expression is:

> "All should/must take responsibility."

This kind of formulation emphasizes the common responsibility for doing something, but in a very general way. However, some respondents are more specific, adding some kind of concrete action, which may vary from the very near 'here and now' (as in: "All should contribute to less garbage and more recycling") to the larger and more complex actions (such as "All must reduce emissions"). In 23 of the occurrences, the common responsibility is emphasized by combining *vi* and *alle* (*we all*).

One of the seven topics induced by the STM analysis was 'International collaboration and responsibilities'. This topic is manifested in different combinations of *alle land* (*all countries*) and highlights the expressed need for both collaboration and commitment:

> "All countries must commit to reducing their emissions."

It also highlights the exemption of the obligations of Norway as a 'small country':

> "All countries should contribute. But [I] deeply disagree that Norway should take the lead and be a pioneer, we are so small [compared to the great powers]."

It is interesting to observe that some refer to the issue of equity, pointing at different responsibilities for rich and poor countries:

> "All countries must take their part of the responsibility, but developed [industrial] countries a larger part of the responsibility than developing countries—for example by helping them to finance climate friendly measures."

The analyses presented here show that there is to a large extent a perception of common and very general responsibilities for solutions to tackle climate change, manifested through the dominant use of *we* and *all*, including the respondents themselves. However, they also point to actors beyond themselves, in particular *the authorities* (37 occurrences), *the state* (73 occurrences), and *the politicians* (83 occurrences). Here are some examples:

> "The authorities must take the lead and facilitate for necessary efforts and necessary measures."

> "The authorities must be tougher and require more, set stricter requirements."

Also, when it comes to *the state* and *the politicians*, the respondents are clear: They should take greater responsibility, govern/make decisions, take the costs, and facilitate for choosing green alternatives. And:

> "The politicians must be willing to adopt unpopular measures."

These answers indicate a critique of the commitment of authorities and politicians as well as lack of and wish for clear decision making.

Deliberation

The analyses presented above may give the impression that the answers are very much aligned to each other, in the sense that the respondents are clear about the fact that something should be done (even though they have different priorities), with the exception of the refusal and hesitation seen among respondents who doubt the influence of human activity on the climate. However, we know that the climate debate in general—due to the comprehensive impact of climate change and the high number of stakeholders involved—is particularly multivoiced or polyphonic. And we do find many traces of deliberation within the answers in the survey discourse under study. Different points of view are brought in, points of view that the respondents have different relations to—e.g., acceptance, agreement, hesitation, or refusal. With a point of departure in results from previous studies (see Fløttum 2010, 2016; Fløttum and Dahl 2011; Fløttum and Gjerstad 2013), I focus here on the polyphony markers *men* (*but*, *however*) and *ikke* (*not*) (see also Chapter 3). The frequency of these markers are 364 occurrences for *men* and 1,038 for *ikke*.

The concessive connective *men*, further referred to as *but*, links two points of view in a more or less clear contrast, which may correspond to different claims. In the construction 'p BUT q', there is no disagreement; the speaker accepts p, but an argumentation is presented, where the q point of view is considered the most important (according to the linguistic instructions embedded in *but*). In addition, the p point of view is oriented toward a conclusion (often implicit) and the q point of view to another one. Let us look at an example where this kind of implicit, or internal, deliberation is presented:

> "[p] It is important to continue as we have started, but [q] it is important to remember that Norway cannot tackle the climate alone."

From this construction we see a point of view, p, leading to an implicit conclusion that 'Norway is doing well', and another point of view, q, leading to the conclusion that 'Norway must collaborate with other countries'. The point of view q is presented as the most important by the respondent, due to its introduction by *but*. Here is another example:

> "[p] Changes at a structural level must be undertaken, but [q] people must take responsibility at the microlevel, also."

The respondent agrees with the point of view p, or concedes that this is a relevant argument, but in this construction he or she considers the point of view q as the main one. In sum, in a construction 'p BUT q', the respondent can agree with both. However, by inserting *but*, the second claim is emphasized as argumentatively stronger than the first one. This type of construction contributes in a subtle way to implicit deliberation and is important to unpack in order to understand how citizens perceive all the possible ways to tackle climate change. The *but* construction in its concessive meaning also indicates how the respondents prioritize between different, and sometimes opposing, points of view. Although all occurrences of *but* in the respondents' answers do not reflect this kind of deliberation, it is notable that this connective appears 364 times.

Another obvious marker of polyphony is the negation adverb *ikke* (*not*), with 1,038 occurrences in our material. Constructions of negation manifested through *not* can convey different meanings. In its polemic use, *not* is clearly polyphonic, integrating two points of view—one underlying point of view and one opposing the underlying point of view—as in this example:

> "The state must be much more active and prohibit 'things'. It does not help that we must pay for polluting—we have enough money."

Here we see that an implicit, underlying point of view—'it helps that we must pay for polluting'—is refuted through the construction with *not*. The contrast is further emphasized by the preceding sentence, claiming more prohibition measures from the state. The respondent is refusing the point of view concerning tax policy, much discussed but approved of in different contexts in Norway, giving an explicit reason after the negated sentence: Norwegians have enough money to pay for such taxes. We also find the 'classic' denial point of view among the negated answers:

> "Climate changes are cyclic and not determined by human beings."

In this example, the underlying point of view about human-caused climate change is refuted. This kind of polyphony is sharper than the one with *but* because it includes an external point of view that the respondent refutes. Through the high frequency of negations, we can infer a high level of polemic deliberation in the material under study. However, note that not all negations are polemic. Examples like the following include negations with a descriptive meaning—i.e., without any obvious underlying point of view:

> "Put up prices (much) on everything that is not climate-friendly."

> "It should cost for those who do not do anything."

In addition, there are approximately 40 occurrences of the answer "Don't know," which has nothing to do with deliberation. It may be an honest answer to the question asked in the survey, but it can also be a refusal of the question itself or an indication of lack of interest.

Conclusion

The purpose of the study presented in this chapter has been to uncover how the public debate has influenced people's engagement in questions related to the transition to a low-carbon society and their willingness to act. The analyses undertaken of the 4,614 answers provided by Norwegian citizens to the open-ended survey question "Concerning climate change, what do you think should be done?" have confirmed the proposed hypothesis: Norwegian citizens are clear about the fact that something should be done to tackle the climate challenge, but they have divergent views on whose responsibilities the proposed actions are. In very general terms, they claim that something must be done and that all should contribute and take responsibility. But many of the survey's respondents do not provide further specifications and thus express only a vague willingness. In the majority of the cases, however, they are specific and enumerate several measures. These answers reflect both national and international political and public debates on solutions to help make the transition to low-carbon development pathways.

There is great variation in the respondents' suggestions and recommendations, and they prioritize in different ways. Prioritizing has to do with choice between different points of view, preferences, and conceptions of the issue. This is clearly manifested in the polyphonic and deliberative answers where different points of view and voices are presented, contrasted, or related to each other by expressions of refusal (such as the negation *not*) or by expressions of contrast and preference (such as the connective *but*).

The semi-automated STM analysis provided seven distinct overall topics that respondents prioritize, including energy transition, emission reduction, development of public transport, changes in lifestyle, governmental measures, international collaboration, as well as the issue of attribution of climate change. Within these, there are several more specific recommendations, varying from the close 'here and now' to the more distant and internationally determined measures. A further finding is that the respondents' answers emphasize mitigation to a much larger extent than adaptation.

The obligation to do something is expressed through a substantial presence of deontic expressions indicating that the state of the world is not what it should be, according to various standards, worldviews, values, and desires, be they political, economic, environmental, or personal. The sentences containing a deontic modal auxiliary generally indicate some action that would change the world so that it becomes closer to an ideal state or to the desired standard. Rhetorically, deontic expressions of obligation convey a clear commanding tone. It is interesting to observe that the commanding tone resulting from the use of deontic modality is sometimes counterbalanced by excuses of this type: "Norway is a small country and cannot do much (on its own)."

Even though there is a dominance of expressions of obligation or recommendation to act in the respondents' answers, the issue of responsibility (Who are the actors? Who should undertake the different suggested actions?) is somewhat vaguely expressed or underspecified, often conveyed through expressions of general, collective responsibility. The respondents point to a common responsibility for solutions to tackle climate change, clearly expressed through the dominant use of *we* and *all*. Together with the accompanying verb expressions, this pronoun use manifests an undeniable willingness of action. The problem here is that, due to the often undefined reference of the pronouns, it is difficult to interpret what/whom these pronouns include (all human beings? people in Norway? everybody on the planet? all institutions?). Linguistically, an expression of obligation presupposes the feasibility of the actions in question and thus, according to a collective implicit voice, that there are actors (humans or institutions) who can effectuate the proposed actions. But who are these actors? The responsibility issue is obviously a difficult one; the responsibilities imply in most cases a division of tasks between authorities (at different levels) and individuals.

It is difficult to interpret to what extent individual respondents include themselves in the 'we' or the 'all'. Some respondents (approximately 200) point at actors beyond themselves, in particular *the authorities*, *the state*, and *the politicians*. In general, these answers demand clearer and more decisive action from governing bodies: The respondents ask for tougher measures, stronger leadership, better facilitation for necessary efforts. It is also interesting to observe that some ask for politicians who dare to work for unpopular and potentially controversial measures. These answers indicate critique of the commitment of the authorities as well as lack of and wish for clear decision making.

The main and general finding of the work reported on here is that citizens have considerable knowledge about mitigation measures and that they are willing to contribute to tackle the challenges, but they ask for better instructions and guidance about how to act individually and for more information about responsibilities. The citizens' answers suggest a willingness to accept stronger mitigation action but claim that authorities and politicians at both local and national levels should facilitate low-carbon choices and thus contribute to bridging policy and individual action.

This study has provided new knowledge on some of the constraints on and opportunities for climate action, which are fundamental to decision making. There is a clear need for more work on clarifying what individuals can do and on distributing responsibilities; a better alignment between decision makers and citizens is required.

Acknowledgments

I am very grateful to my colleagues Øyvind Gjerstad, Mikael Poul Johannesson, Åsta Dyrnes Nordø and Endre Tvinnereim for the collaboration on the Structural Topic Modelling (STM) part of the research reported on in this chapter.

References

Aldy, J. E., M. J. Kotchen, and A. A. Leiserowitz. 2012. "Willingness to pay and political support for a US national clean energy standard." *Nature Climate Change* 2: 596–599.

Anthony, L. 2011. "AntConc: Design and development of a freeware corpus analysis toolkit for the technical writing classroom." *Proceedings of Professional Communication Conference*, 729–737.

Faarlund, J. T., S. Lie, and K. I. Vannebo. 1997. *Norsk referansegrammatikk*. Oslo: Universitetsforlaget.

Fløttum, K. 2010. "A linguistic and discursive view on climate change discourse." *Asp-Revue du GERAS* 58: 19–37.

Fløttum, K. 2016. "Linguistic analysis approaches for assessing climate change communication." In: Von Storch, H. (ed.). *Climate science: Oxford research encyclopedias*. Oxford: Oxford University Press. doi: 10.1093/acrefore/9780190228620.013.488

Fløttum, K., and T. Dahl. 2011. "Climate change discourse: Scientific claims in a policy setting." *Fachsprache* 3–4: 205–219.

Fløttum, K., and T. Dahl. 2014. "IPCC communicative practices: A linguistic comparison of the summary for policymakers 2007 and 2013." *The LSP Journal—Language for Special Purposes, Professional Communication, Knowledge Management and Cognition* 5 (2): 66–83.

Fløttum, K., T. Dahl, and V. Rivenes. 2016. "Young Norwegians and their views on climate change and the future: Findings from a climate concerned and oil rich nation." *Journal of Youth Studies* 19 (8): 1128–1143. http://dx.doi.org/10.1080/13676261.2016.1145633

Fløttum, K., A. M. Gjesdal, Ø. Gjerstad, N. Koteyko, and A. Salway. 2014. "Representations of the future in English language blogs on climate change." *Global Environmental Change* 29: 213–222.

Fløttum, K., and Ø. Gjerstad. 2013. "Arguing for climate policy through the linguistic construction of narratives and voices: The case of the South-African green paper 'National Climate Change Response'." *Climatic Change* 118 (2): 417–430. doi: 10.1007/s10584-012-0654-7

Gellein, M. L., G. Kipperberg, and A. V. Risa. 2015. "Preferanser for norske utslippsreduksjoner versus klimatilpasningstiltak." *Samfunnsøkonomen* 3: 18–32.

Hulme, M. 2009. *Why we disagree about climate change*. Cambridge: Cambridge University Press.

Kerbrat-Orecchioni, C. 2002. "Présupposé, présupposition." In: Charaudeau, P., and D. Maingueneau (eds.). *Dictionnaire d'analyse du discours*, pp. 467–469. Paris: Seuil.

Kotchen, M. J., K. J. Boyle, and A. A. Leiserowitz. 2013. "Willngness-to-pay and policy—instrument choice for climate-change policy in the United States." *Energy Policy* 55: 617–625.

Krosnick, J. A., and B. MacInnis. 2013. "Does the American public support legislation to reduce greenhouse gas emissions?" *Daedalus* 142: 26–39.

Leiserowitz, A. 2005. "American risk perceptions: Is climate change dangerous?" *Risk Analysis* 25: 1433–1442.

Lorenzoni, I., A. Leiserowitz, M. de Franca Doria, W. Poortinga, and N. F. Pidgeon. 2006. "Cross-national comparisons of image associations with 'global warming' and 'climate change' among laypeople in the United States of America and Great Britain." *Journal of Risk Research* 9: 265–281.

Lorenzoni, I., and N. Pidgeon. 2006. "Public views on climate change: European and USA perspectives." *Climatic Change* 77: 73–95.

Lyons, J. 1977. *Semantics*. Cambridge: Cambridge University Press.

Moser, S. C. 2010. "Communicating climate change: History, challenges, process and future directions." *WIREs Climate Change* 1: 31–53.

Nerlich, B., N. Koteyko, and B. Brown. 2010. "Theory and language of climate change communication." *WIREs Climate Change* 1: 97–110.

Nølke, H., K. Fløttum, and C. Norén. 2004. *ScaPoLine. La théorie scandinave de la polyphonie linguistique*. Paris: Kimé.

Obradovich, N., and S. Guenther. 2016. "Collective responsibility amplifies mitigation behaviors." *Climatic Change* 137: 307–319.

Quirk, R., S. Greenbaum, G. Leech, and J. Svartvik. 1972. *A grammar of contemporary English*. London: Longman.

Roberts, M. E., B. M. Stewart, D. Tingley, C. Lucas, J. Leder-Luis, S. K. Gadarian, et al. 2014. "Structural topic models for open-ended survey responses." *American Journal of Political Science* 58: 1064–1082.

Smith, N., and A. Leiserowitz. 2014. "The rise of global warming skepticism: Exploring affective image associations in the United States over time." *Risk Analysis* 32: 1021–1032.

Tvinnereim, E., and K. Fløttum. 2015. "Explaining topical prevalence in open-ended survey questions on climate change." *Nature Climate Change* 5: 744–747. doi: 10.1038/nclimate2663

Tvinnereim, E., K. Fløttum, Ø. Gjerstad, M. P. Johannessen, and Å. D. Nordmo. Forthcoming. "Citizens' preferences for tackling climate change: Quantitative and qualitative analyses of their freely formulated solutions."

Tvinnereim, E., and E. Ivarsflaten. 2016. "Fossil fuels, employment, and support for climate policies." *Energy Policy* 96: 364–371. doi: http://dx.doi.org/10.1016/j.enpol.2016.05.052

Tvinnereim, E., and S. I. Steinshamn. 2016. "Folkelig aksept for klima- og energitiltak i Norge." *Samfunnsøkonomen* 130: 77–86.

8 The Paris COP21 Agreement—Obligations for 195 Countries

Kjersti Fløttum and Helge Drange

Introduction

The Conference of the Parties (COP), which is the governing body of the United Nations Framework Convention on Climate Change (UNFCCC), has for several years strived for an international agreement on the climate change challenge. This became particularly important after the COP15 summit in Copenhagen in 2009 (UNFCCC 2010), which generally was characterized as a complete failure. Six years later, the international community concluded at COP21, December 12, 2015, the Paris Agreement, signed by 195 countries (i.e., the Parties), stating the aim to "strengthen the global response to the threat of climate change." This was applauded as a historical event in responding to the challenges represented by global climate change. However, at the same time, it has been characterized as of no value without each country's commitment to it. The Paris Agreement is not legally binding; and for a successful achievement, it is heavily dependent on each country's Intended Nationally Determined Contributions (INDC—and, after the onset of the agreement, NDC).

Since the Paris Agreement was signed, there has been a lot of discussion about the importance or value of the agreement in both mainstream and social media as well as in published reports and papers. In the literature published during the first half of 2016, different perspectives are taken, and the discussion often turns around the question to what extent the agreement can be assessed as a success and a historic breakthrough or as a discouraging failure. Obergassel *et al.* (2016, 3) label the Paris Agreement a "success" despite shortcomings such as lack of legal constraints. They affirm that the question is not "whether the Paris Agreement will deliver the emission reductions necessary, but whether the agreement has the potential to catalyze further changes, whether it becomes a pacemaker for policy processes—at the international level and in the capitals of the world. From this perspective, the Paris Agreement is much stronger than many had expected." However, regarding language use, in particular related to financing, they consider it as "vague" concerning "concrete financing contributions for mitigation and adaptation in developing countries" (Obergassel *et al.* 2016, 47).

In another perspective, Savaresi (2016, 16) assesses the outcome of the conference against the expectations, concluding on a reflection about "where the climate regime is likely to head next, and where the Paris Agreement leaves Parties in their efforts to tackle climate change." She does not see the Paris Agreement as "a miraculous cure for all the maladies of the climate regime" but believes the Parties can undertake the hard work lying ahead "in a considerably much [more] hopeful mood, certainly the most hopeful it has been for quite some time" (Savaresi 2016, 28). In a paper focusing on the role of nonstate actors in tackling climate change, Asselt (2016, 91) claims that they can make a valuable contribution by holding "the parties to their obligations under the Paris Agreement." In another and more pessimistic perspective, Althor, Watson, and Fuller (2016, 20281) take as a point of departure the inequity in the causes and effects of climate change ("Global mismatch between greenhouse gas emissions and the burden of climate change") and conclude that, without a binding agreement established at COP21 for INDCs, "there is no clear indication of how successful the Paris Agreement will be." Starting with a rather negative judgment ("The December 2015 Paris Climate Agreement is better than no agreement. This is perhaps the best that can be said about it"), Clémençon (2016, 3) seems to conclude that more climate action "will not occur without determined and far-reaching government intervention in energy markets in the next few years, particularly in the largest polluting countries."

The document itself may be seen as a general action guide for the 195 countries, including a number of procedures and measures that require strict attention, accounting, transparency, and numerous follow-ups. Thus, some crucial questions can be raised: How does the document integrate scientific findings that the requested action is expected to be based on? And what is the rhetorical tone that should encourage the different Parties to commit to the measures and instruments in question?

Before investigating and discussing these questions, we will provide some general characteristics of the document. The document published in Paris, December 12, 2015, at the closure of the Conference of the Parties COP21, consists of 32 pages (UNFCCC 2015). The first part (pages 1–20) concerns the adoption of the agreement and the decisions to give effect to the agreement; the second part (Annex) constitutes the agreement itself (pages 21–32, approximately 7,340 words), named PARIS AGREEMENT. It consists of a one-page introduction, constituting the context recognized as the background for the agreement, followed by 29 articles (Art), most of them consisting of several paragraphs. This second part—the agreement—will be the main focus in this chapter. Except for a range of articles concerning various formalities and administrative issues (Art1, 15–29), Art2–14 are devoted to topics directly related to the global response to the threat of climate change: aim (Art2), nationally determined contributions (Art3), mitigation/reduction of greenhouse gas emissions (Art4), sinks and reservoirs of greenhouse gases / forests (Art5), mixed mitigation / adaptation (Art6), adaptation

(Art7), loss and danger (Art8), financial support (Art9), technology development and transfer (Art10), capacity building (Art11), education (Art12), transparency framework (Art13), and stocktaking (Art14).

The divergent assessments that are found in existing literature on the issue, demonstrating different perspectives, constitute an interesting point of departure for the analysis that we will undertake in the present chapter. In the next section, we will consider the presentation and interpretation of scientific findings in the agreement. Then, in the subsequent section, we take on a more linguistic perspective in an analysis of the choice of words and their combinations in order to reveal the overall rhetorical tone of the document. To the best of our knowledge, little has been done concerning linguistic analysis going beyond the mere description of language use as "clear" or "vague." Our hypothesis is that the rhetorical tone reflects the necessity of political action, based on recent scientific conclusions, but in a sufficiently general way to obtain the necessary approval by the various national assemblies. We will sum up the findings in our "Concluding Remarks" section.

Science in the Paris Agreement

The Paris Agreement has, as expected, limited reference to and no discussion of the underlying science. The science is rather indirectly acknowledged by statements including words and phrases like "Acknowledging," "Emphasizing," "Recognizing," "Taking full account of," and "in accordance with best available science"—in a few cases with explicit references but generally with implied references to the latest reports of the IPCC and previous COPs under the auspices of the UNFCCC. Regarding its form, the Paris Agreement represents a continuation of some of the previous amendments—e.g., from COP16 in Cancún, 2010 (UNFCCC 2011).

In the Annex of the Paris Agreement, only one physical property—"Global average temperature"—and three weather/climate phenomena or events are mentioned, with the latter group expressed in rather general terms: "Extreme weather events," "Slow onset events," and "Events that may involve irreversible and permanent loss and damage." Of these, only "global average temperature" is explicitly quantified, and then in terms of the upper, allowable (human-induced) warming by the end of the 21st century relative to preindustrial temperatures. This "climate target" or "temperature goal" is thus the concretization of the overarching precautionary principle of the UNFCCC as agreed upon at the Earth Summit in Rio de Janeiro in 1992: to "prevent dangerous anthropogenic interference with the climate system" (UNFCCC 1992, §2). A common interpretation of the precautionary principle is that human-induced climate changes should not threaten food production, violate sustainable development of nations, or deteriorate ecosystems (e.g., Oppenheimer and Petsonk 2005).

Although not stated, one can assume that "global mean temperature" is the global mean SURFACE temperature, including both ocean surface and land surface temperatures. Also worth pointing out is that the interpretation of "dangerous climate" cannot be resolved by science in isolation but requires a blend of political, economic, ethical, scientific, cultural, and other considerations (e.g., Knutti *et al.* 2015; Moss 1995; O'Neill and Oppenheimer 2002; SED 2015). Science can, however, develop the basis for and present sets of mitigation measures required to meet a given climate target based on present knowledge (e.g., Parry *et al.* 1996; SED 2015).

Global mean surface temperature is undeniably a convenient proxy and a practical metric for identifying human-induced changes in climate since the global surface temperature responds to altered forcing of the climate system, and since other changes in climate often (albeit with important exceptions; see below) can be related to the global mean surface temperature. Importantly, surface temperature recordings based on thermometers extend back to the second half of the 19th century for major parts of the globe, enabling quantification of an approximate preindustrial reference value. In addition, surface temperature can be communicated and understood—at least in general terms—by non-experts across cultures and geographical boarders. The above does not imply that the global mean surface temperature is THE optimal metric describing the state of climate (e.g., Bopp *et al.* 2013; Knutti *et al.* 2015; Oppenheimer and Petsonk 2005; Steinacher, Joos, and Stocker 2013) or that a specific temperature goal is absolute in the meaning that any warming in excess of the goal immediately triggers large-scale and/or dangerous, abrupt, or irreversible changes to climate. Rather, it is clearly one of the available, workable, understandable, and operational metrics around.

The "temperature goal" in the Paris Agreement is based on the widely known and much discussed two-degree target, for the first time formally adopted at COP16 in Cancún (UNFCCC 2011). To the surprise of many, the Parties at the Paris meeting strengthened the temperature target from Cancún in two ways. First, the temperature target at COP21 states that the increase in global average temperature should be "well below 2 °C," in contrast to the "below two degrees" from COP16. The Paris Agreement does not concretize the meaning of "well below." An indication is, however, provided by the additional intention to "pursue efforts to limit the temperature increase to 1.5 °C." A natural interpretation of the Paris Agreement is thus that measures should be taken to ensure that the increase in global average (ocean and land surface) temperature by the end of the century relative to the preindustrial value should be as close to 1.5 °C as possible.

Whereas the two-degree target has been discussed for several decades in scientific reports and in the peer-reviewed literature (Collins *et al.* 2013; Jaeger and Jeager 2011; Oppenheimer and Petsonk 2005; Randalls 2010; Rijsberman and Swart 1990; Rogner *et al.* 2007: 99–100; WBGU 1995) and from the mid- to late 1990s gradually implemented in politics (e.g.,

European Union Council 1996: item no. 6), the 1.5 °C target—albeit hardly found in the scientific literature as a numerically hard-defined target—has emerged as a response to the (possible) inadequacy of the two-degree target. The latter has particularly been based on concerns in respect to extreme weather events, water availability, crop yields, coral reef degradation, and sea-level rise (e.g., Fawcett *et al.* 2015; Hansen and Sato 2004; Hansen *et al.* 2016; Harvey 2007; Oppenheimer *et al.* 2014).

The Paris Agreement's endorsement of the 1.5 °C target seemingly includes a paradox. First, the first part of the agreement acknowledges the need for additional research regarding the target (paragraph 21, page 4):

> (The Conference of the Parties [. . .]) "*Invites* the Intergovernmental Panel on Climate Change to provide a special report in 2018 on the impacts of global warming of 1.5 °C above pre-industrial levels and related global greenhouse gas emission pathways."

At the same time, Art4a states that the 1.5 °C target "would significantly reduce the risks and impacts of climate change." The latter is partially supported by science (e.g., Hansen and Sato 2004; Harvey 2007, Oppenheimer *et al.* 2014). However, as summarized by Schleussner *et al.* (2016), "comprehensive, multi-sectoral assessments of differences in climate impacts between a 1.5 °C and 2 °C warming are lacking." In addition, the lowest emission scenario adopted for the IPCC AR5 is in line with the two-degree target (with 66% probability; Collins *et al.* 2013), preventing a thorough assessment of the stricter target.

The prominent position of the stricter 1.5 °C target is likely a result of repeatedly raised concerns from the Alliance of Small Island States (AOSIS) and the least-developed countries (LDCs) that the two-degree climate target is inadequate to prevent—particularly for the group of nations mentioned—dangerous climate change (Tschakert 2015). Whereas the summary from COP13 in Bali in 2007 states that "an increase of 2 °C in global temperature would have devastating impacts on small island developing states" (Earth Negotiations Bulletin 2007), all of the summaries from COP14 (2008) to COP20 (2014) include explicit requests from AOSIS and LDCs to limit the warming to maximum 1.5 °C (see the respective *Earth Negotiations Bulletins* at http://www.iisd.ca/enb/vol12/).

For AOSIS, the concern has been particularly related to one of the most unavoidable consequence of global warming: increasing sea level. For these states, higher sea level is, in the utmost consequence, an existential threat to their people, history, and heritage (e.g., Barnett and Adger 2003; Gibbons and Nicholls 2006; Marzeion and Levermann 2014; McNamar and Gibson 2009; Nicholls 2004; Woodroffe 2008). A consequence of this, together with general concerns about the ability to adapt to changes in climate for the LDCs, lead a total of 106 nations—corresponding to more than two thirds of the Parties—to declare their support for limiting the warming to maximum 1.5 °C prior to COP21 (Manilla Communiquè 2015). With this backdrop, it is hard to see how an agreement in Paris could be made without incorporation of the 1.5 °C temperature target.

Limiting global warming to "well below 2 °C" requires quick and substantial reductions in the total human-induced emissions of greenhouse gasses (Rogelj *et al.* 2016a, 2016b; Seneviratne *et al.* 2016). Quantification of the required emission reduction is, however, not given in the Paris Agreement, although Section II of the first part of the agreement documents that the INDCs are clearly insufficient to meet the two-degree target. The degree of necessary emission cuts is vaguely described in Art4.1, which states the need to "achieve a balance between anthropogenic emissions by sources and removals by sinks of greenhouse gases in the second half of this century." This statement, that anthropogenic greenhouse gas emissions need to be balanced by sinks in the Earth system, dubbed the "net zero emissions" target by Geden (2016), can be viewed as the third "climate target" endorsed by the Parties at COP21. Hidden in the latter target—and consistent with the lowest emission scenario used in IPCC AR5 (Collins *et al.* 2013)—is the possible need for capturing CO_2 from the atmosphere, artificially or by means of enhanced photosynthesis on land and in the ocean, followed by safe and long-term sequestration of the captured carbon. The option for so-called "negative emissions" (Fuss *et al.* 2014; Gasser *et al.* 2015) is thus embedded in the Paris Agreement.

It is still unclear whether negative-emission technologies are technically, economically, environmentally, or financially viable on the scale needed for the accumulated CO_2 emissions not to overshoot the two-degree and 1.5-degree targets (Anderson and Peters 2016). Anderson and Peters (2016) phrase the negative-emission technologies implicitly embedded into the Paris agreement as a "high-stakes gamble" and state that deep reliance on these are a "moral hazard par excellence."

It is well recognized that nations—depending on geographic location, political system, tradition, national income, technological advancement, historical emissions, standard of living, etc.—have very different opportunities regarding adapting to climate change and to initiate and maintain a diverse suite of mitigation measures. The Paris Agreement acknowledges these challenges by defining, together with the "climate target" (Art1a), adaptation to climate change (Art1b) and financial means for mitigation and adaptation (Art1c) as the three foundations for the agreement.

Lexical Choices and Rhetorical Tone

In order to provide a well-founded analysis of the document, it is important to clarify what kind of text genre it is. It may appear as a diplomatic or even judicial text with its different articles indicating actions to be undertaken, listed one after another. However, it is not a legal document, even though it is binding to the countries (Parties) in the sense that it calls on them to follow up on the different instruments and measures enumerated. Another interesting characteristic is that it is directed at the different Parties (i.e., countries), and at the same time it is the countries who are "speaking." On the first page, it says: "The Parties to this Agreement [. . .] Have agreed as follows:"—followed by Articles 1–29. This makes a mix of sender and receiver, which marks the document a kind of "self-instruction."

Since it is an agreement with instructions and approved procedures, we expect to find a clear presence of deontic expressions—i.e., expressions with meanings related to obligations, often expressed through modal verbs such as *ought*, *shall*, *should*, and *must* (Le Querler 1996; Quirk *et al.* 1972). This marks a change in language use and communication about climate change, which, through the Assessment Reports published regularly by IPCC, has been dominated by epistemic expressions related to (un)certainty and probability, degrees of confidence, evidence, and agreement, mostly strictly predefined according to fixed scales established by the IPCC (Fløttum and Dahl 2014). This is due to the fact that these reports are based on scientific findings and to the role of the IPCC as being policy relevant but not policy descriptive. However, in the Summary for Policy Makers of the IPCC Synthesis Report published in 2014, there is a clear emphasis on risk perspectives (Fløttum, Gasper, and StClair 2016). Because of this emphasis, and the fact that the Parties recognize "the need for an effective and progressive response to the urgent threat of climate change on the basis of the best available scientific knowledge" (p. 21), one would expect forceful language use on the necessity to undertake actions—rapidly, locally, and globally—in order to tackle climate change. In the following, this will be investigated, mainly through a study of the verbs used in combination with modal auxiliary verbs. This choice is justified by the evident fact that verbs represent the word class denoting action.

Use of Modal Auxiliary Verbs

A modal auxiliary verb is a type of verb that is used to indicate modality—i.e., a quality or modification added to the main verb of the sentence, such as likelihood, (un)certainty, ability, permission, and obligation. Some modal verbs (e.g., *must*) can have different meanings, depending on the context. In English the most common auxiliaries are *can/could*, *may/might*, *must*, *will/would*, and *shall/should* (see Quirk *et al.* 1972). The frequency of modal auxiliaries provides an impression of the rhetorical tone in a document. In Table 8.1, the frequency of such modals in the Paris Agreement is listed.

We observe that the dominating auxiliary in the agreement is *shall*, represented by the highest frequency—117 occurrences. *Shall* is generally an infrequent English auxiliary with restricted use compared to other auxiliaries such as *should*, *will*, and *would*, but in this context it is used in a "quasi-legal" sense and with insistence ("strong volition") (see Quirk *et al.* 1972, 99). We will come back to this below. The second-most frequent auxiliary

Table 8.1 Frequency of Modal Auxiliary Verbs

Auxiliary Verb	*Frequency*
shall	117
should	25
may	19
will	6
can	4
would	1
could	0
might	0
must	0

in the agreement is *should* (25 occurrences). This auxiliary has usually a milder meaning than *shall*, expressing a guideline or a recommendation, but that can be oriented toward an obligation or a necessity (similar to *ought to*, which is not used in the present document). We find 19 occurrences of *may*, which usually expresses a permission to do something or is used to talk about a possible and not certain situation. The other auxiliaries used are represented by modest frequencies—*will* (6) expresses some prediction on the future, which can be specific (with temporal expression) or non-specific (timeless); *can* (4) expresses ability, permission, or possibility; and *would* (1), which is less definite than *will*, has many different uses, including invitations, requests, asking permission, and talking about preferences (see Murphy 2012).

From Table 8.1, we also see that the auxiliary verb *must*—with a strong deontic meaning, denoting a clear obligation or compulsion—is not used, which may be regarded as somewhat surprising, given the urgency of the issue at stake, as recognized by the signing Parties in the introduction to the articles of the agreement: "the need for an effective and progressive response to the urgent threat of climate change on the basis of the best available scientific knowledge" (p. 21). Unsurprisingly, the weaker modals *could* and *might* are not used: *could* expressing less definite aspects of possibility or suggested options than *can*, and *might*, not always clearly different from *may*, but used when the chances of something being the case are unlikely or hypothetical.

Words and Expressions Co-occurring with the Auxiliaries

From these observations, we conclude, as a first rhetorical characteristic of the Paris Agreement, that it is dominated by a tone of command or exhortation through the frequent use of the auxiliary *shall*, often used in laws, regulations, or directives to express what is mandatory (see Merriam-Webster Online Dictionary). However, since an auxiliary's purpose is to help forming the mode (and tense) of other verbs, and in that sense does not have any definite meaning on its own, it is necessary to investigate which verbs the auxiliaries used co-occur with. These verbs and verb expressions are enumerated alphabetically in Table 8.2, with number of occurrences in parentheses (taken into account here is the verb immediately following *shall*).

Table 8.2 Frequency of Verbs Co-occurring with Modal Auxiliary

Modal Auxiliary Verb—N	*Construction Type*	*Verb / Verb Expression*
SHALL 117	Active *shall* (78)	account for (1), adopt (2), aim (3), apply (7), build on (1), collaborate (1), communicate (3), consider (2), consist of (2), cooperate (1), decide (1), declare (1), do (1), engage (1), ensure (1), enter into force (2), establish (1), exercise (2), not exercise (1), form (2), have (1), identify (1), include (1), inform (4), keep under review (1), make (1), notify (1), operate (1), participate (1), pay attention to (2), perform (1), prepare (1), promote (2), provide (6), pursue (1), reflect (1), recognize (1), serve (as) (6), specify (1), strengthen (1), take effect (1), take into account/consideration (3), take stock of (1), undergo (1), undertake (1)
	Passive *shall be* (30) *shall not be* (3)	applied (1), authorized (1), bound (1), considered (1), convened (1), deposited (2), enhanced (1), held (3), open for (3), provided (for) (5), recognized (1), recorded (2), replaced (2), restricted (1), submitted (1), supervised (1), supported (1), taken (2) counted (1), entitled to (1), used to (1)
	shall be— followed by other word classes (6)	expert (1), responsible for (2), subject to (2), the Depository (1).
SHOULD 25	Active *should* (21)	aim (1), communicate (1), continue (3), cooperate (1), enhance (3), facilitate (1), follow (1), provide (3), represent (1), respect (1), strengthen (1), strive (1), submit (1), take action to (1), take into account (1)
	Passive *should be* (3)	based on (1), country-driven (1), guided (1)
	should be— followed by other word classes (1)	an effective, iterative process (1)

Modal Auxiliary Verb—N	*Construction Type*	*Verb / Verb Expression*
WILL 6	Active *will* (5)	allow (1), benefit (1), represent (2), take longer (1)
	Passive *will be* (1)	implemented
WOULD 1	Active (1)	reduce
CAN 4	Active *can* (3)	contribute (1), involve (1), reduce (1)
	Passive *can be* (1)	used
MAY 19	Active *may* (10)	adjust (1), include (3), involve (1), participate (2), prepare (1), provide (1), withdraw (1)
	Passive *may be* (9)	admitted (1), affected (1), deemed (1), decided (1), enhanced (1), made (1), represented (1), required (1), specified (1)

We will not go into detail about all the attested combinations; but the complete lists of verbs should provide a valuable overview. We will focus on the most frequent auxiliary—*shall*—and the verbs that it co-occurs with. First, without taking into consideration the difference between active and passive constructions, we observe that there is a major presence of verbs denoting some administrative or formal procedure: *account for*, *adopt*, *apply*, *authorize*, *decide*, *declare*, *deposit*, *exercise*, *keep under review*, *notify*, *pay attention to*, *record*, *submit*, *take into account/consideration*, *take stock of*. These are not what we usually associate with action—i.e., actions that should contribute to reaching the aim of reducing greenhouse gas emissions, adapting to ongoing and forthcoming changes in climate, and introducing mitigation measures. The verbs used bear a sense of bureaucracy, connected with many complicated procedures and ways of doing things. In fact, the verb *reduce* occurs only three times and only once in the combination "reduce greenhouse gas emissions" (Art10, 1; in the two others, *reduce* co-occurs with "the risks and impacts of climate change" and "the need for additional adaptation efforts"). This observation points to the responsibility of each Party to concretize and implement the obligations they have agreed to fulfill, a responsibility that is further emphasized in the objects of the verbs used. Here are some examples (our emphasis):

> "Parties shall account for **their nationally determined contributions.**" (Art4, 13)
>
> "The outcome of the global stocktake shall inform **Parties in updating [. . .] their actions and support.**" (Art14, 3)

"The Conference of the Parties serving as the meeting of the Parties to the Paris Agreement shall keep under regular review **the implementation of this Agreement.**" (Art16, 4)

"Developed country Parties shall provide **transparent and consistent information** on support for developing country Parties." (Art9, 7)

"Each Party should, as appropriate, submit and update periodically **an adaptation communication.**" (Art26, 10)

In addition to the dominant formal and bureaucratic tone, the agreement also emphasizes the importance of cooperation, represented through verbs such as *collaborate*, *communicate*, *cooperate*, *participate*. Article 12 enumerates areas related to education and awareness as areas where cooperation is required:

> "Parties shall cooperate in taking measures, as appropriate, to enhance climate change education, training, public awareness, public participation and public access to information, recognizing the importance of these steps with respect to enhancing actions under this Agreement." (Art12)

Actors and Beneficiaries

The main actors are the Parties themselves, as stated in the introduction to the agreement, and in some cases the Conference of the Parties (as illustrated in the examples above). However, given the emphasis in the introduction on "the principle of equity and common but differentiated responsibilities and respective capacities," it is interesting to examine how this is followed up through distinguishing between different Parties in the articles. The distinction between developed and developing countries is well taken care of through the specifications "developed country Parties" (9 occurrences), "developing country Parties" (38 occurrences), and, further, "the least developed countries and small island developing States" (4 occurrences). In the following example, we also observe that "particularly vulnerable" is included:

> "Parties recognize that adaptation is a global challenge faced by all with local, subnational, national, regional and international dimensions, and that it is a key component of and makes a contribution to the long-term global response to climate change to protect people, livelihoods and ecosystems, taking into account the urgent and immediate needs of those developing country Parties that are particularly vulnerable to the adverse effects of climate change." (Art7, 2)

The phrase "particularly vulnerable to the adverse effects of climate change" co-occurs in fact 6 times with "developing country Parties" and illustrates that the latter are not only actors but also beneficiaries to the extent that they, as "particularly vulnerable," will be provided support from the other Parties. However, as indicated by Obergassel *et al.* (2016), the financial support aimed at for developing countries is not expressed in a particularly strict or clear way:

Article 9

1 Developed country Parties shall provide financial resources to assist developing country Parties with respect to both mitigation and adaptation in continuation of their existing obligations under the Convention.
2 Other Parties are encouraged to provide or continue to provide such support voluntarily.

Even though the majority of the sentences with *shall* (78) have an active form, 33 of the total of 117 have a passive form. The passive construction can be a rhetorical way of not making explicit who is the agent or actor of the expressed action, as in the following example:

> "Support shall also be provided for the building of transparency-related capacity of developing country Parties on a continuous basis."
> (Art13, 15)

The responsible actor for providing the support described in this sentence is not made explicit, and thus this construction contributes to a vagueness of the agreement, even though who the beneficiaries are is expressed ("developing country parties"). However, most of the passive constructions introduced by *shall be* include some kind of actor, as "by the secretariat" in the following:

> "Nationally determined contributions communicated by Parties shall be recorded in a public registry maintained by the secretariat."
> (Art4, 12)

Another way of avoiding to express who or what should be the responsible actor(s) for the required actions expressed by a verb is the phenomenon of nominalization. Nominalization is a process where a part of speech (most often a verb or an adjective) is converted into a noun, as from *reduce* to *reduction*. In this process, the actor can be omitted: "The reduction of greenhouse gases is important in order to tackle climate change." As mentioned above, the verb *reduce* is used only three times in the agreement, while *reduction* occurs six times, and they are in fact mostly co-occurring with the expression of actors, as in the following:

> "In order to achieve the long-term temperature goal set out in Article 2, Parties aim to reach global peaking of greenhouse gas emissions as soon as possible, recognizing that peaking will take longer for developing

> country Parties, and to undertake rapid reductions thereafter in accordance with best available science."
>
> (Art4, 1)

However, with, for example, the nominalization *mitigation*, which is represented by 23 occurrences in the agreement, another question may be raised. Although the actors are explicitly present, the object of mitigation—i.e., in which sectors mitigation must take place—is not stated:

> "Each Party shall prepare, communicate and maintain successive nationally determined contributions that it intends to achieve. Parties shall pursue domestic mitigation measures, with the aim of achieving the objectives of such contributions."
>
> (Art4, 2)

This is another reminder of the responsibility of the Parties to find and exploit the best and most effective "mitigation measures" in order to attain the objectives stated by the agreement.

Quasi-Absence of People

Climate change and its effects is something that has impacts on people and their lives and lifestyles. However, people are surprisingly absent from the agreement (see also Fløttum, Gasper, and StClair 2016) except for this statement from the introduction:

> "Acknowledging that climate change is a common concern of humankind, Parties should, when taking action to address climate change, respect, promote and consider their respective obligations on human rights, the right to health, the rights of indigenous peoples, local communities, migrants, children, persons with disabilities and people in vulnerable situations and the right to development, as well as gender equality, empowerment of women and intergenerational equity."

In the articles, we find only three additional mentions of *people*, and none of *children* (perhaps the most vulnerable of all), *migrants*, or *persons*. Interestingly, the word *gender* appears twice in the construction *gender-responsive*, as in the following statement:

> "Parties acknowledge that adaptation action should follow a country-driven, gender-responsive, participatory and fully transparent approach."
>
> (Art7, 5)

People have lifestyles, and different lifestyles according to economic situation and culture, a matter that is extremely relevant to questions about tackling climate change. However, we find the word *lifestyles* only once, in the introduction, where the following is recognized:

> "[S]ustainable lifestyles and sustainable patterns of consumption and production, with developed country Parties taking the lead, play an important role in addressing climate change."
>
> (Introduction, p. 21)

Urgency

In the introduction to the agreement's articles, the Parties recognize (our emphasis)

> "the need for an effective and progressive response to the **urgent** threat of climate change on the basis of the best available scientific knowledge."
>
> (Introduction, p.21)

Given the circumstances and the intense discussions both before and during the COP21 meeting, one would expect a clear insistence on urgency in the document. However, there is only one additional mention:

> "taking into account the **urgent** and immediate needs of those developing country Parties that are particularly vulnerable to the adverse effects of climate change."

Although we find one occurrence of the time adverbial *as soon as possible*, we notice that many other expressions indicate a non-emergency tone, such as the following: *as appropriate* (15 occurrences), *over time* (2 occurrences), *to the extent possible* (1 occurrence), *voluntary/-ily* (6 occurrences). These expressions represent a contrast to the verbal expressions introduced by the forceful *shall*.

Our final remark on the use of language in the agreement relates to its generally positive tone. Despite the overall commanding tone conveyed through expressions of obligation, statements like the following contribute to a positive tone through the verbs *promote*, *enhance*, and *enable*:

> [The proposed approaches] shall aim to:
>
> (a) Promote mitigation and adaptation ambition;
> (b) Enhance public and private sector participation in the implementation of nationally determined contributions; and
> (c) Enable opportunities for coordination across instruments and relevant institutional arrangements.
>
> (Art6, 8)

Variants of *enhance/enhancement* reach the frequency of 30 occurrences. These and other positive-laden words (such as the choice of *contribution* in the *nationally determined contributions—NDC*) as well as the urge for cooperation corroborate the point of view of Savaresi (2016, 28) that with this agreement the Parties can undertake the hard work lying ahead "in a considerably much hopeful mood, certainly the most hopeful it has been for quite some time."

Concluding Remarks

The analysis reported in this chapter has investigated the representation and interpretation of scientific findings and the rhetorical tone in the Paris Agreement adopted December 12, 2015, at the Conference of the Parties (COP21).

The science in the Paris Agreement is generally presented by implicit reference to the latest reports of the IPCC (and to the amendments from the previous COPs) by statements starting with "Acknowledging," "Emphasizing," "Recognizing," and "Taking full account of." The overarching precautionary principle of the UNFCCC—to "prevent dangerous anthropogenic interference with the climate science"—is concretized in the agreement by two temperature targets: first, that the global mean surface temperature should be "well below 2 °C" and, second, that efforts should be made "to limit the temperature increase to 1.5 °C." Neither the type of target (in this case, surface temperature) nor the setting of thresholds on the target (2 °C or 1.5 °C) to fulfill the UNFCCC principle can be made by science in isolation. The climate targets in the agreement should therefore be considered as the net result of political, economic, ethical, scientific, and cultural considerations made by the Parties, including building on the outcome of the statements and agreements from the preceding COPs. With respect to the latter, the widely acclaimed "success" of the Paris Agreement could possibly not have been achieved without the generally viewed "failure" of, for instance, COP15 in Copenhagen 2009 and COP16 in Cancún in 2010. Actually, large parts of the Paris Agreement follow closely the amendments from preceding COPs, notably from the COP16, demonstrating the evolutionary process toward the Paris Agreement.

The lexically based linguistic analysis of the agreement has enabled us to confirm the hypothesis that the rhetorical tone reflects the necessity of political action to tackle climate change. However, the urgency of the issue seems remarkably under-communicated, even though there is a strongly expressed will to fulfill the aim of strengthening the global response to the threat of climate change. This will is mainly conveyed through expressions including the auxiliary *shall*, through denoting obligation, and through a bureaucratic style making the rhetorical tone very formal. The choice of the auxiliary *shall* in a sentence affects the intensity of a requirement—orders are given. Furthermore, the agreement calls for action by referring to instruments and measures in a general and abstract way. The tone is thus sufficiently general to be approved by the 195 Parties, avoiding going into details on more specific requirements where interests may be very divergent and opinions controversial. The signing Parties have agreed to impose these requirements on themselves and will need to undertake substantial and urgent actions (the Nationally Determined Contributions) to be concretized and implemented in their respective countries. Status and progress on both national and global scales will furthermore be reviewed and updated every five years. However, the rhetorical tone is also positive, not least through the urging for cooperation, and that developed countries have a wider and more pressing responsibility than less developed countries. This positive flavor, if maintained, may help the Parties to consider the tackling of climate change more as an opportunity than as a costly threat.

References

Althor, G., J. E. M. Watson, and R. A. Fuller. 2016. "Global mismatch between greenhouse gas emissions and the burden of climate change." *Scientific Reports* 6: 20281. doi: 10.1038/srep20281. http://www.ncbi.nlm.nih.gov/pmc/articles/PMC4742864/

Anderson, K., and G. Paters. 2016. "The trouble with negative emissions." *Science* 354: 182–183. doi: 10.1126/science.aah4567

Asselt, H. van. 2016. "The role of non-state actors in reviewing ambition, implementation, and compliance under the Paris agreement." *Climate Law* 6 (1–2): 91–108. doi: 10.1163/18786561-00601006

Barnett, J., and W. N. Adger. 2003. "Climate dangers and atoll countries." *Climatic Change* 61: 321–337. doi: 10.1023/B:CLIM.0000004559.08755.88

Bopp, L., L. Resplandy, J. C. Orr, S. C. Doney, J. P. Dunne, M. Gehlen, et al. 2013. "Multiple stressors of ocean ecosystems in the 21st century: Projections with CMIP5 models." *Biogeosciences* 10: 6225–6245. doi: 10.5194/bg-10-6225-2013

Clémençon, R. 2016. "The two sides of the Paris climate agreement: Dismal failure or historic breakthrough?" *Journal of Environment & Development* 25 (1): 3–24. doi: 10.1177/1070496516631362

Collins, M., R. Knutti, J. Arblaster, J.-L. Dufresne, T. Fichefet, P. Friedlingstein, X. et al. 2013. "Long-term climate change: Projections, commitments and irreversibility." In: Stocker, T. F., D. Qin, G.-K. Plattner, M. Tignor, S. K.Allen, J. Boschung, et al. (eds.). *Climate change 2013: The physical science basis: Contribution of working group I to the fifth assessment report of the intergovernmental panel on climate change*, pp. 1045–1152. Cambridge & New York: Cambridge University Press.

Earth Negotiations Bulletin 2007. Summary of the Thirteenth Conference of Parties to the UN Framework Convention on Climate Change and Third Meeting of Parties to the Kyoto Protocol, December 3–15. http://www.iisd.ca/climate/cop13/. Accessed 11 July 2016.

European Union Council. 1996. *Community strategy on climate change—Council conclusions*. 1939th council meeting environment, document Nr. 8518/96. http://ue.eu.int/ueDocs/cms_ Data/docs/pressData/en/envir/011a0006.htm

Fawcett, Allen A., G. Iyer, L. E. Clarke, J. A. Edmonds, N. E. Hultman, H. C. McJeon, et al. 2015. "Can Paris pledges avert severe climate change?" *Science* 350: 1168–1169. doi: 10.1126/science.aad5761

Fløttum, K., and T. Dahl. 2014. "IPCC communicative practices: A linguistic comparison of the summary for policymakers 2007 and 2013." *The LSP Journal—Language for Special Purposes, Professional Communication, Knowledge Management and Cognition* 5 (2): 66–83.

Fløttum, K., D. Gasper, and A. L. StClair. 2016. "Synthesizing a policy-relevant perspective from the three IPCC 'Worlds'—a comparison of topics and frames in the SPMs of the fifth assessment report." *Global Environmental Change* 38: 118–129. doi: 10.1016/j.gloenvcha.2016.03.007

Fuss, S., J. G. Canadell, G. P. Peters, M. Tavoni, R. M. Andrew, P. Ciais, et al. 2014. "Betting on negative emissions." *Nature Climate Change* 4: 850–853. doi: 10.1038/nclimate2392

Gasser, T., C. Guivarch, K. Tachiiri, C. D. Jones, and P. Ciais. 2015. "Negative emissions physically needed to keep global warming below 2°C." *Nature Communications* 6, 1–7. doi: 10.1038/ncomms8958

Geden, O. 2016. "An actionable climate target." *Nature Geosci* 9: 340–342.

Gibbons, S. J. A., and R. J. Nicholls. 2006. "Island abandonment and sea-level rise: An historical analog from the Chesapeake Bay, USA." *Global Environmental Change* 16: 40–47.

Hansen, J., and M. Sato. 2004. "Greenhouse gas growth rates." *Proceedings National Academy Sciences* 101: 16109–16114. doi: 10.1073/pnas.0406982101

Hansen, J., M. Sato, P. Hearty, R. Ruedy, M. Kelley, V. Masson-Delmotte, et al. 2016. "Ice melt, sea level rise and superstorms: Evidence from paleoclimate data, climate modeling, and modern observations that 2°C global warming could be dangerous." *Atmospheric Chemistry and Physics* 16: 3761–3812. doi: 10.5194/acp-16–3761–2016

Harvey, L. D. D. 2007. "Dangerous anthropogenic interference, dangerous climatic change, and harmful climatic change: Non-trivial distinctions with significant policy implications." *Climatic Change* 82 (1): 1–25. doi: 10.1007/s10584-006-9183-6

Jaeger, C. C., and J. Jaeger. 2011. "Three views of two degrees." *Regional Environmental Change* 11: 15–26. doi: 10.1007/s10113-010-0190-9

Knutti, R., J. Rogelj, J. Sedlácek, and E. M. Fischer. 2015. "A scientific critique of the two —degree climate change target." *Nature Geoscience* 9: 13–19. doi: 10.1038/NGEO2595

Le Querler, N. 1996. *Typologie des modalités*. Caen: Presses universitaires de Caen.

The Manilla Communicé. 2015. Outcome statement of the Sherpa Senior Officers' Meeting of the Climate Vulnerable Forum (CVF), Manila, November 9–11. http://www.thecvf.org/wp-content/uploads/2015/11/THE-MANILA-COMMUNIQUE6.pdf. Downloaded 11 July 2016.

Marzeion, B., and A. Levermann. 2014. "Loss of cultural world heritage and currently inhabited places to sea level rise." *Environmental Research Letters* 9:1–7. doi: 10.1088/1748–9326/9/3/034001

McNamara, K. E., and C. R. Gibson. 2009. " 'We do not want to leave our land': Pacific ambassadors at the United Nations resist the category of 'climate refugees'." *Geoforum* 40: 475–483. http://dx.doi.org/10.1016/j.geoforum.2009.03.006

Merriam-Webster Online Dictionary. http://www.merriam-webster.com/

Moss, R. H. 1995. "Avoiding 'dangerous' interference in the climate system." *Global Environmental Change* 5: 3–6.

Murphy, R. 2012. *English grammar in use*. Cambridge: Cambridge University press.

Nicholls, R. J. 2004. "Coastal flooding and wetland loss in the 21st century: Changes under the SRES climate and socio-economic scenarios." *Global Environmental Change* 14: 69–86.

Obergassel, W., Chr. Arens, L. Hermwille, N. Kreibich, F. Mersmann, H. E. Ott, and H. Wang-Helmreich. 2016. *Phoenix from the ashes—An analysis of the Paris agreement to the United Nations framework convention on climate change*. Wuppertal: Wuppertal Institute for Climate, Environment and Energy.

O'Neill, B., and M. Oppenheimer. 2002. "Dangerous climate impacts and the Kyoto Protocol." *Science* 296: 1971–1972.

Oppenheimer, M., M. Campos, R. Warren, J. Birkmann, G. Luber, B. O'Neill, and K. Takahashi. 2014. "Emergent risks and key vulnerabilities." In: Field, C. B., V. R. Barros, D. J. Dokken, K. J. Mach, M. D. Mastrandrea, T. E. Bilir, M. et al. (eds.). *Climate change 2014: Impacts, adaptation, and vulnerability. Part A: global and sectoral aspects. contribution of working group II to the fifth assessment report of the intergovernmental panel on climate change*, pp. 1039–1099. Cambridge & New York: Cambridge University Press.

Oppenheimer, M., and A. Petsonk. 2005. "Article 2 of the UNFCCC: Historical origins, recent interpretations." *Climatic Change* 73: 195–226. doi: 10.1007/s10584-005-0434-8

Parry, M. L., Carter, T. R., and Hulme, M. (1996). "'What is a dangerous climate change?" *Global Environmental Change* 6 (1): 1–6

Quirk, R., S. Greenbaum, G. Leech, and J. Svartvik. 1972. *A grammar of contemporary English*. London: Longman

Randalls, S. 2010. "History of the 2C climate target." *WIREs Climate Change* 1: 598–605. doi: 10.1002/wcc.62

Rijsberman, F., and R. Swart. 1990. *Targets and indicators of climate change. Report of working group II of the advisory group on greenhouse gases*. Stockholm: Stockholm Environmental Institute.

Rogelj, J., M. Schaeffer, P. Friedlingstein, N. P. Gillett, D. P. van Vuuren, K. Riahi, et al. 2016a. "Differences between carbon budget estimates unravelled." *Nature Climate Change* 6: 245–252. doi: 10.1038/NCLIMATE2868

Rogelj, J., M. den Elzen, N. Höhne, T. Fransen, H. Fekete, H. Winkler, R. Schaeffer, F. Sha, K. Riahi, and M. Meinshausen 2016b. "Paris agreement climate proposals need a boost to keep warming well below 2°C." *Nature* 534: 631–639. doi: 10.1038/nature18307

Rogner, H.-H., D. Zhou, R. Bradley, P. Crabbé, O. Edenhofer, B. Hare (Australia), et al. 2007. *Introduction*. In: Metz, B., O. R. Davidson, P. R. Bosch, R. Dave, and L. A. Meyer (eds.). *Climate change 2007: Mitigation*, pp.95–116. Contribution of Working Group III to the Fourth Assessment Report of the Intergovernmental Panel on Climate Change Cambridge & New York: Cambridge University Press.

Savaresi, A. 2016. "The Paris agreement: A new beginning?" *Journal of Energy & Natural Resources Law* 34 (1): 16–26. doi: 10.1080/02646811.2016.1133983

Schleussner, C.-F., T. K. Lissner, E. M. Fischer, J. Wohland, M. Perrette, A. Golly, et al. 2016. "Differential climate impacts for policy-relevant limits to global warming: The case of 1.5 °C and 2 °C." *Earth System Dynamics* 7: 327–351. doi: 10.5194/esd-7–327–2016

SED: UNFCCC: Report on the Structured Expert Dialogue (SED) on the 2013–2015 review, FCCC/SB/2015/INF.1, 2015.

Seneviratne, S. I., M. Donat, A. J. Pitman, R. Knutti, and R. L. Wilby. 2016. "Allowable CO2 emissions based on regional and impact-related climate targets." *Nature* 529: 477–483. doi: 10.1038/nature16542

Steinacher, M., F. Joos, and T. F. Stocker. 2013. "Allowable carbon emissions lowered by multiple climate targets." *Nature* 499: 197–201. doi: 10.1038/nature12269

Tschakert, P. 2015. "1.5°C or 2°C: A conduit's view from the science policy interface at COP20 in Lima, Peru." *Climate Change Responses* 2: 1–11. doi: 10.1186/s40665–015–0010-z

UNFCCC. 1992. *United Nations framework convention on climate change*. Report No. FCCC/INFORMAL/84. https://unfccc.int/resource/docs/convkp/conveng.pdf. Downloaded 12 July 2016.

UNFCCC. 2010: Decision 2/CP.15, Copenhagen Accord (30 March 2010). http://unfccc.int/resource/docs/2009/cop15/eng/11a01.pdf. Downloaded 12 July 2016.

UNFCCC. 2011: Decision 1/CP.16, The Cancun Agreements (15 March 2011). http://unfccc.int/resource/docs/2010/cop16/eng/07a01.pdf. Downloaded 12 July 2016.

UNFCCC. 2015: Draft decision -/CP.21, Adoption of the Paris Agreement (12 December 2015). https://unfccc.int/resource/docs/2015/cop21/eng/l09r01.pdf. Downloaded 12 July 2016.

WBGU (German Advisory Council on Global Change). 1995. *Scenario for the derivation of global* CO_2 *reduction targets and implementation strategies: Statement on the occasion of the first conference of the parties to the framework convention on climate change in Berlin*. Bremerhaven: Special Report, WBGU.

Woodroffe, C. D. 2008. "Reef-island topography and the vulnerability of atolls to sea-level rise." *Global and Planetary Change* 62: 77–96.

New Methods to Study the Language of Climate Change

9 Data-Driven Approaches to Climate Change Discourse, Illustrated through Case Studies of Blogs and International Climate Negotiation

Andrew Salway

Introduction

Observations about the increasing volume and diversity of textual material available for humanities and social science research are at the tipping point of cliché but nevertheless point to exciting and challenging opportunities. The climate change debate manifests in a great diversity of texts—in multiple genres and many languages across the globe—produced by a wide variety of actors concerned with different aspects of climate change and wishing to promote different points of view. In addition to scientific reports, policy documents, and newspaper stories that have been quite well studied, recent review papers highlight the need to pay more attention to social media data such as blogs and microblogs like Twitter (Auer, Zhang, and Lee 2014; Schäfer 2012; Schäfer and Schlichting 2014). On top of already large amounts of news media, the widespread uptake of social media means that the volume of textual material relating to climate change is vast and increasing rapidly. Recent collections of English-language texts alone suggest 30,000 newspaper articles, 2.8 million tweets, and 150,000 blog posts per month broadly related to climate change (Olteanu *et al.* 2015; Salway *et al.* 2016).

Compare these numbers to the sample sizes that are feasible when manual methods are used to investigate climate change discourse—for example: manually coding frames in 974 newspaper stories (Boykoff 2008); identifying 8 distinct "discourses" from about 150 newspaper stories (Doulton and Brown 2009); analyzing key themes in 3 blogs based on the reading of 20 blog posts from each (Sharman 2014); and analyzing frames in 300 tweets (O'Neill *et al.* 2015). It is apparent that automatic text analysis techniques must be incorporated into research methods in order to deal with the volume of climate change texts. Of course, it is crucial to consider the trade-off between the depth, richness, and sensitivity of analysis that is possible for a human analyst and the breadth of coverage that automated techniques enable. At the very least, however, automation has a role in identifying texts that are related to the topic of climate change: e.g., the use of queries to identify relevant texts for investigating the coverage of climate change in news media and social media, both over time and geographically (Boykoff and Roberts 2007; Grundmann and Scott 2014; Kirilenko, Molodtsova, and Stepchenkova 2015; Olteanu *et al.* 2015; Schmidt, Ivanova, and Schäfer 2013). Beyond that, the focus in this chapter is on the use of automated techniques to analyze the content of textual data sets, known as corpora, related to climate change—i.e., to provide manageable overviews of the content, to highlight interesting linguistic patterning, and to automatically annotate, or code, the material. I argue that the scale, complexity, variety, and dynamic nature of climate change discourses mean that data-driven approaches in particular have an important role in analyzing large text corpora. In brief, data-driven approaches may help to minimize biases due to prior assumptions about the form and content of the textual material and help to generate new insights and stimulate hypotheses.

The next section discusses what "data-driven" means specifically in relation to automatic text analysis, and it sketches a general approach for data-driven research. This is followed by a brief review of the text analysis techniques that have been used to investigate climate change discourse, especially in news media and social media—i.e., text classification, topic modeling, and various corpus linguistic techniques that have been used for corpus-based discourse analysis. The review concludes that corpus-based discourse analysis is important in order to understand texts as more than "bags of words"—i.e., to not only treat language at the lexical level, and that new techniques are required to enable researchers to identify linguistic patterning in large corpora. These points are then explored and developed through three case studies that reflect on the data-driven text analysis techniques that were incorporated into corpus-based methods for investigating climate change discourse. The latter two case studies made use of a recently developed technique for elucidating linguistic patterning—local grammar induction—which is introduced in its own section.

What Does "Data-Driven" Mean for Automatic Text Analysis?

Automatic text analysis, in essence, involves counting instances of linguistic forms that have meaning with respect to a conceptual framework that, in turn, articulates aspects of phenomena that are the object of an investigation—e.g., topics, media representations, frames, rhetorical strategies, and ideologies. A simple case would be the counting of words associated with different topics in order to analyze the topics' prevalence across a corpus of texts. Thus, when considering the relative merits of different approaches to text analysis, we should pay attention to how decisions are made in relation to (i) selecting non-textual phenomena for investigation; (ii) determining an appropriate conceptual framework; (iii) establishing a mapping between concepts and linguistic forms that can be counted (i.e., a coding scheme); and (iv) identifying interesting and significant statistical results in the frequency counts of linguistic forms.

A premise of the current chapter is that the increasing volume and diversity of textual material related to climate change mean that data-driven approaches will become increasingly relevant, and perhaps essential, as part of each of (i) to (iv) above, and especially (iii). The diversity of material from different domains, comprising various text genres, and in multiple languages, means that the situation in which researchers are faced with textual material that they are not familiar with will become commonplace. Thus, it will be harder for researchers to rely on previously developed theory, conceptual frameworks, and coding schemes to guide and implement their investigations. As for volume, when text corpora already comprise hundreds and thousands of millions of words, even if a researcher is only interested in one particular word or term, then the researcher may be faced with hundreds of thousands of relevant sentences, which precludes close reading of much of the material. As well as meeting a practical need for automated analysis, there may also be a methodological preference for data-driven approaches that minimize prior assumptions about the form and content of the textual material and thus provide less biased, or at least alternatively biased, views on the data that can challenge and extend current theory.

Scientists have been using data-driven approaches for centuries. That is to say, they have at times developed theory based on masses of observations, rather than through hypothesis-driven experimentation. Take, for example, the astronomical Rudolphine Tables that Kepler used to derive laws of planetary motion, and Darwin's extensive and detailed descriptions of animal and plant life that led to the theory of evolution. In more recent times, "discovery science"—now based on automatically detected patterning in vast electronic data sets—has come to be a norm in some branches of science, such as astronomy, genomics, and neuroscience. A leading neuroscientist, when considering the relative status and merits of discovery science and hypothesis-driven research, wrote with great candor: "If we were to be truthful, we would admit that often hypotheses are retrofitted to our accidental observations. Moreover, these un-hypothesized observations are, in many cases, more exciting and revealing that anything we would have ever posited. They spur us in directions that no pure thought process ever would. If we are very lucky, they lead us to revise many of the hypotheses previously conceived" (Akil 2003, 4).

Importantly, it is apparent that in these statements data-driven discovery science is seen not as an alternative to hypothesis-driven research but rather as a stage prior to the formulation and testing of hypotheses. Indeed, the choice of exclusive alternatives that is implied in the distinction between "data-driven" and "hypothesis-driven" is unnatural and unhelpful—likewise, the parallel distinction between "deductive" and "inductive" research methods, and the "top-down versus bottom-up" metaphor where theory is placed above data. Working deductively, researchers start with a theory that suggests hypotheses to be tested against data generated from an experiment. Working inductively, researchers start with data about that part of the world (or universe) in which they are interested, before detecting patterns and regularities, the explanations for which lead to the formulation of new hypotheses and theory. Of course, in practice, it is not possible to begin selecting data, let alone interpreting it, without some input from prior theory either in the mind of the researcher or manifested in the computer program used to analyze the data. However, when working more inductively, in an explorative mode, then a researcher is perhaps more likely to make observations that challenge and extend existing theory. There is of course also the risk of wandering fruitlessly with no clear objective through masses of data on a "fishing trip" (a term often used to disparage data-driven research).

Thus, in the context of automatic text analysis, I am interested in what have been termed "discovery tools" (Shneiderman 2001): These are tools that combine data mining and visualization techniques so that researchers can explore alternative views on large data sets. There is always a human in the loop to direct the analysis and, of course, to interpret results and theorize. This resonates with the point that automatic content analysis methods are best for "amplifying and augmenting careful reading and thoughtful analysis" (Grimmer and Stewart 2013, 268). Envisage a researcher faced with a large volume of textual material with which he or she is not very familiar. Discovery tools should elucidate salient and idiosyncratic linguistic forms in the corpus that could inform decisions about the previously mentioned steps (i) to (iv), as follows.

(i) Selecting non-textual phenomena to investigate. Of course, the researcher likely has some phenomena in mind already when choosing to analyze a particular text corpus, but discovery tools may draw attention to linguistic forms that prompt interest in other phenomena too.

(ii) Determining an appropriate conceptual framework. Again, the researcher will probably have a conceptual framework in mind at the outset, but this may be confirmed or revised by reading the corpus with discovery tools, especially with regard to (iii).

(iii) Establishing a mapping between concepts and linguistic forms that can be counted—i.e., developing a coding scheme. The crux of automatic text analysis is the counting of linguistic forms—words, classes of words, phrases, grammatical structures, etc.—that map to the concepts and non-linguistic phenomena that are pertinent to a particular investigation. If this mapping is established prior to analyzing the corpus, then there is a risk of the "seek and ye shall find" problem. That is, if my coding scheme, or my intuition, says that concept X manifests in linguistic forms A, B, and C, then I will never know if there are other ways that X commonly manifests in the corpus or if there are other related phenomena and concepts in the corpus that should be considered. A data-driven approach would first identify salient linguistic forms in the corpus. Then some of these forms would be mapped to phenomena and concepts through manual interpretation by confirming or modifying an existing conceptual framework, or by developing one from scratch. This gives a more rigorous and comprehensive mapping between the conceptual framework and the corpus at hand. Of course, this depends on relevant phenomena manifesting in sufficiently frequent and regular linguistic forms such that the forms can be identified with automatic techniques. These techniques can only access the text in contrast with a human coder who can, for better or worse, draw on context and general world knowledge. This places a fundamental limit on the applicability of purely data-driven text analysis techniques.

(iv) Identifying interesting and significant statistical results in the frequency counts of linguistic forms. Once frequency counts of linguistic forms have been generated, then the research is akin to other sciences that work with numeric data. So I will not spend time considering this phase here, except to note that the research may continue in either a data-driven or a hypothesis-driven manner and may return to iterate through the earlier steps in the light of the results that are generated.

Given the great diversity of textual material related to climate change, then techniques to elucidate salient and idiosyncratic linguistic forms in a corpus should be applicable and reliable across different languages, text genres, and domains. This means that the techniques should have a minimal reliance on prior linguistic descriptions that some techniques depend upon, such as lexicons (vocabularies) and grammars. Such resources are not available for most languages, text genres, and domains, and they are costly to make on a case-by-case basis. Furthermore, these resources can introduce implicit biases into research. Given that linguists cannot even agree on apparently straightforward descriptions such as parts of speech (to wit, multiple part-of-speech tagging schemes, each delimiting a different number of parts of speech), let alone a theory of semantics, then the prospect for computationally tractable linguistic descriptions that are generally applicable seems remote, if not a de facto impossibility. In other words, it seems wise, as far as possible, to let each corpus speak for itself—or "trust the text" (Sinclair 2004)—so that the idiosyncrasies of different domains, text genres, and languages are not lost under generic schemes. To sum up, there are practical and methodological reasons to prefer text analysis techniques that do not depend upon previously created linguistic resources that encode linguistic descriptions and assumptions that cannot be generally applicable.

A Brief Review of Automatic Text Analysis Techniques That Have Been Used to Investigate Climate Change Discourse

Broadly, there are two kinds of ways in which automated techniques have been used to investigate climate change discourse, although they are not mutually exclusive: (i) automated techniques have been used to code texts on a greater scale than is possible manually (e.g., by classifying each text in terms of subtopics or frames); (ii) automated techniques have been used under the rubric of corpus-based discourse analysis to identify and interpret distinctive linguistic patterning.

For automatic coding, typically the text is taken as the principle unit of analysis, and each text is treated as a bag of words—i.e., at solely the lexical level such that information about word order and higher-level linguistic structures are ignored. This is commonly the basis for text classification, text clustering, and text scaling; see Grimmer and Stewart (2013) for an in-depth description and discussion of these techniques. Classification techniques assume that classes are known in advance. Often a set of manually classified exemplar texts is used to train an automatic classifier; however, this does not seem to have been used for climate change texts. Otherwise, classification proceeds on the basis of combinations of words to define classes of texts; this is similar to the earlier mentioned use of queries to retrieve texts when investigating news media and social media coverage. The mapping between words and classes of texts may be specified a priori or may be developed inductively.

In a simple case, hashtags were used to identify and count tweets associated with different aspects of climate change discourse (Kirilenko and Stepchenkova 2014); in other work, different themes induced from a set of hashtags were used to categorize tweets (Pearce *et al.* 2014). Six kinds of climate change "narrative" were induced from the transcripts of TV news broadcasts and mapped to detailed combinations of words that define each kind of narrative and enable automated search for them (Mayer 2012). For example, part of the query for the "climate tragedy" narrative states that either the term "global warming" or the term "climate change" should occur in the proximity of a word such as "Katrina," "drought," "tornado," "flood," etc.; another part of the query seeks to exclude stories denying climate change by stating that stories should not include words such as "hoax," "alarmist," etc. The categorization was based on "extensive reading of media coverage, as well as conversations with journalists and observers of the media" (5), and the combinations of words were developed by testing "a variety of keyword search strings to find a formula that most consistently identified stories that fit each narrative" (41). A similar approach was taken to analyze climate change frames in Twitter (Jang and Hart 2015), although the combinations of words to identify each frame were somewhat simpler—e.g., a climate change tweet was considered to be an instance of the "hoax frame" if it contained "hoax," "lie," or "fraud." The relevant words were selected from the most frequent words for the top 500 tweets (ranked by the amount they were retweeted).

In the studies mentioned above, establishing the mappings between words and classes was somewhat ad hoc in each case. More systematic analyses may be achieved by using unsupervised techniques for scaling and clustering, such as topic modeling, which are data driven in that scales and clusters are induced automatically from the texts (which are typically treated as bags of words). Then, insights are gained both from where documents are placed in clusters and on scales, and from inspection of the words that are most discriminative. Tvinnereim and Fløttum (2015) used structural topic modeling to induce four topics from 2,115 open-ended survey responses in order to characterize public opinion about climate change and to identify the most representative responses of each topic for close reading. The advantage of unsupervised techniques in general is that they obviate the cost, and possible biases, of manually labeling some texts as training examples that must be done for supervised techniques. A disadvantage is that sometimes the techniques must be tuned—i.e., to set parameters for the best results, which may mean that many differing outputs must be validated or discounted by the researcher.

It has been established that the processing of texts as bags of words is "sufficient to infer substantively interesting properties of texts (Hopkins and King 2010)" (Grimmer and Stewart 2013, 273). However it is important to recognize that multiple linguistic structures—including many that arise from the sequencing of words in terms, phrases, grammatical constructions, etc.—contribute to the meaning of a text. When treating a text as a bag of words, there is a risk of losing most of the meaning arising from these structures, so it is important to consider what other aspects of meaning can be elucidated using data-driven techniques. Thus, I now turn to corpus-based discourse analysis—i.e., the second kind of way in which automated text analysis techniques have been used to investigate climate change discourse.

Corpus linguistics has well-established techniques for exploring linguistic patterning in large text corpora in a data-driven manner while ensuring that at least some information about word sequences is kept. Common techniques include frequency lists (giving the frequency of each word), keywords (words that are unusually frequent in a particular corpus), concordances (a set of text fragments around a word of interest), word clusters / n-grams (frequent word sequences including the word of interest), and collocations (words tending to occur with a word of interest); for an introduction to these techniques, see McEnery and Hardie (2012). In recent years, there has been considerable interest in applying these techniques under the rubric of corpus-based discourse analysis in order to "uncover linguistic patterns which can enable us to make sense of the ways that language is used in the construction of *discourses* (or ways of constructing reality)" (Baker 2006, 1): A crucial point is that much linguistic patterning would not be apparent to the researcher without the use of automated techniques. The approach is exemplified in the work of Baker *et al.* (2008), who used collocation and concordance analysis to identify common categories of media representations (related to refugees, asylum seekers, immigrants, and migrants) and representative texts for qualitative analysis from a 140-million-word corpus of newspaper articles. Similar approaches (variously using frequency lists, word clusters, collocates, and concordances) have been used to investigate climate discourses—e.g., concordances of word clusters to analyze "carbon compounds" in web RSS feeds (Koteyko 2010); collocates of frequent words to get a view of the framing of climate change in newspapers (Grundmann and Krishnamurthy 2010); the collocates of environmental terms to determine their use in positive, negative, and neutral contexts (Wild *et al.* 2013); and a temporal and geographic comparison of keywords and word clusters in newspapers to identify the main claim makers and the relative visibility of acceptors and skeptics (Grundmann and Scott 2014).

In general, frequent words, keywords, and word clusters can be seen as highlighting salient concepts in discourses. However, these techniques can only be a starting point for a researcher. A lack of information about the co-text around keywords and word clusters restricts the extent to which they can be interpreted without the close reading of concordances. Increasingly, corpora are too large for close reading of enough relevant concordances, so automated techniques are needed to condense information about co-texts—i.e., about the sequences of words around a word of interest. Collocation data gives a view on which words tend to co-occur with the word of interest and hence can be seen as giving insights into the meanings typically associated with it. However, collocation data is typically presented as a large grid of statistics for one keyword, and it would be desirable to have a simpler picture that is more intuitive to interpret. Also, while collocation data describes a relationship between two lexical items, it does not capture further sequential information. A step toward considering the sequencing of words when analyzing frames has been taken by researchers using text visualization techniques to explore word associations around a term of interest (Baumer *et al.* 2013; Diakopoulos, Zhang, and Salway 2013; Diakopoulos *et al.* 2014, 2015). Toward a similar aim, the method for local grammar induction described later in this chapter is intended to provide a condensed view of the patterning in words sequences around a word of interest. First, though, the next section presents a case study to reflect on the use of some established corpus linguistic techniques as part of a data-driven approach.

Case Study I: Inducing Categories of Future Representations from Climate Change Blogs

Fløttum *et al.* (2014) investigated representations of the future in climate change discourse in the blogosphere, using a frequency list, word clusters, sorted concordances, and also simple template patterns. At the outset of this work, the interest in representations of the future was established, but little was known and nothing assumed about how such representations might be manifested in the corpus to be analyzed. The NTAP corpus comprises about 3,000 English language blogs (1.4 million blog posts) related to climate change issues from 2000–2012 (Salway, Hofland, and Touileb 2013): From this, all 330,000 sentences containing either "climate change" or "global warming" were selected for analysis.

Our first step was to generate a frequency list and inspect the 1,500 most frequent words in order to identify those that we judged might form part of representations of the future—e.g., the word "future" itself, and others including "threat," "risk," "danger," and "opportunity." Limiting ourselves to the top 1,500 words was an arbitrary but expedient decision, in keeping with the idea of looking for the most salient linguistic forms. The 11 selected words had some 30,000 instances in total, so there was still too much data for it to be feasible to read and analyze all of the relevant co-texts. Instead, word clusters and sorted concordances were generated in order to elucidate patterning around the selected words. Frequency-ordered lists of word clusters showed the most common sequences containing each word of interest—e.g., "future of" and "future for." Sorted concordances presented text snippets around the words of interest, sorted alphabetically according to the surrounding words: They were useful to reveal patterns such as "a WORD future," where WORD can stand for any word (see Figure 9.1).

Working in this way, 42 patterns related to representations of the future were identified—three of these patterns are shown in Table 9.1. Each pattern contains a filler (i.e., 'WORD'); in some patterns, there is alternation that is shown with the '|' symbol and optionality shown with brackets. For each selected pattern, a program counted the number of instances of the pattern, the number of different fillers appearing in the pattern, and the frequency of each filler; the table shows the top five fillers. This data provided a condensed view of the co-texts around the 11 selected words that reveals common ways in which the words are used in the corpus.

Inspection of this data for all 42 patterns, along with the close reading of co-texts for certain pattern-filler combinations, led to the identification of nine

future that will awaken humanity to a green future we need to strengthen our team to push
will unleash a wave of investment in a green future," he told delegates at the opening
about opportunities in investing in a green future "insiders place bets on global warming
needs to work with china to build a greener future on a foundation of coal if we are
opportunity from its commitment to a greener future and the fight against global warming
to change their habits if offered a happy future to look forward to rather than a bleak
climate change and creating a healthier future for generations to come
56-page handbook, sustaining a healthy future – taking action on climate change
protect the environment, and ensure a healthy future for our families
and dreams to save the planet from a hellish future I consider the global warming
as a "potent warning rather than a hopeful future", and he is completely right in
of climate change and want to have a livable future I simply cannot see why climate
catastrophic climate change and defend a livable future the success of earth hour is
won't just provide the planet a living future, it actually will create far more jobs

Figure 9.1 Part of a Sorted Concordance for the Word "future," Which Reveals the Pattern "a WORD future"; Similar to That Used by Fløttum *et al.* (2014)

Table 9.1 Three of the Patterns Derived from Manual Inspection of Word Clusters and Sorted Concordances by Fløttum *et al.* (2014)

Pattern	*Unique fillers*	*Total Instances*	*Number of Instances for the Five Most Frequent Fillers*
a|an WORD future	97	239	sustainable (34); low-carbon (19); better (15); uncertain (12); greener (7)
risk(s)|danger(s)|threat(s) facing WORD	30	142	the (43); our (26); humanity (25); mankind (10); humankind (5)
opportunit(y|ies) to WORD	325	843	make (39); address (18); put (16); build (16); take (15)

meaning categories to characterize the different future representations in the corpus: (1) sustainability, (2) value-laden positive, (3) value-laden negative, (4) temporal, (5) future for people/human beings and future of humanity/planet, (6) future for and of regions/countries, (7) future for nature/environment, (8) future for business/industry/economy, (9) future for security. This categorization provided the basis for further investigations—e.g., to test the hypotheses that "accepting" climate change blogs would be more concerned with the future than "skeptical" blogs, and that, when they did address the future, then the "skeptical" blogs would typically be responding to the discourses of the "accepting" blogs (Salway, Fløttum, and Elgesem 2015). The established mapping between the categories and common ways in which they are realized—i.e., certain pattern-filler combinations—meant that it was relatively straightforward to make quantitative comparisons between the two corpora and to identify samples of future representations in each sub-corpus for close reading.

This work exemplifies the idea of using data-driven techniques to provide a manageable view over a large text corpus and, in concert with manual interpretation of the results and close reading of samples, to assist in establishing a conceptual framework and a mapping between concepts and countable linguistic forms. It should be noted that, in this example, the mapping between categories and their textual realizations is not comprehensive—rather, as the result of a frequency-led analysis, we expect that it captures the most common textual realizations. Furthermore, we cannot guarantee that every instance of a certain linguistic form (pattern-filler combinations in this case) is being used to convey the same meaning—we assume that most of them are, based on the close reading of some examples. The method that identified salient patterns around frequent words and provided a condensed view of their co-texts (e.g., the examples in Table 9.1) relied on manual analysis of lists of word clusters and sorted concordances, which was somewhat ad hoc and time consuming. In the following section, I present and discuss a recently developed technique that generates similar kinds of data about the co-texts around selected words in a wholly automated way.

Local Grammar Induction: A New Technique to Induce Salient Information Structures for Data-Driven Content Analysis

This technique is intended to induce salient information structures from unannotated corpora as a discovery tool for humanities and social science research. It preserves information about word order in text so that more linguistic structure, and hence meaning, is elucidated than with bag-of-words approaches. Since it does not rely on any linguistic resources, other than a corpus, it should be portable and relatively free of biases. Preliminary results suggest that the technique provides a condensed view of selected words' co-texts that may be useful when analyzing topics, framing, rhetorical strategies, and relations between entities (see case studies II and III).

The technique, described in detail in Salway and Touileb (2014), is based on ADIOS (Solan *et al.* 2005), which is an unsupervised grammar induction algorithm that induces hierarchical structures from sequential data—e.g., words in the sentences of a corpus. Using statistical information, it identifies the most significant word sequences, referred to here as H(orizontal)-groups and equivalence classes—V(ertical)-groups. H-groups and V-groups may be nested within others. For presentation, the sequential items in an H-group are separated by whitespace, and the alternative items in a V-group are separated by the '|' symbol, such that the output reads like a simple regular expression. For example, take "((climate change)|(global warming)) is caused by." Here the top level H-group comprises a sequence starting with a V-group, which itself contains two alternative H-groups, and then three words. Statistical evidence from a corpus has led the technique to recognize that "climate change" and "global warming" are common word sequences and that they both often precede "is caused by."

The ADIOS algorithm, like other grammar induction algorithms, builds on the insights of Zellig Harris, who argued that grammatical structures can be induced through a distributional analysis of the surface forms of languages (Harris 1954). He also showed how linguistic structures that are identified in this way map to important information structures, especially in domain-specific corpora (Harris 1988). This second point motivated our work to modify and apply the ADIOS algorithm for data-driven text analysis. One important modification was to focus the algorithm on text snippets around one single key term of interest at a time, rather than processing all the sentences in a corpus. This change was influenced by the theory of local grammar (Gross 1997)—i.e., the idea that language is best described with word classes and combinations that are specific to the local contexts of individual words.

Figure 9.2 shows 10 of the 671 top-level H-groups that were induced by Salway and Touileb (2014) from the NTAP corpus (the corpus is described in case study I); the technique was applied separately to 17 key terms ("climate change," "sea levels," "carbon tax," etc.), which ranged in frequency from thousands to hundreds of thousands. Each of the 10 top-level H-groups in Figure 9.2 contributes something to an overall impression of what is commonly written in the co-texts of key terms in the corpus. H-groups 1, 2, and 3 each capture alternatives for core domain terminology—e.g., H-group 2 highlights the alternatives "anthropogenic global warming," "manmade global warming," and "man made global warming." Likewise, there are alternatives in 4, but these seem to reflect different strengths of feeling about the need to deal with climate change—i.e., "combat," "minimize," or "tackle." H-group 5 captures four different ways to express that something is caused by climate change or global warming. The next two H-groups are included to show that not all of the output is meaningful: It seems that 6 is incomplete, and 7 has formed a structure with no basis in language. The remaining examples (8–10) include some errors but nevertheless give a summary impression of common statements—e.g., "pollution blamed for global warming," "should regulate greenhouse gases," and "climate models suggest that," and some of their alternative phrasings in the corpus.

Of course, the examples presented here are in no way surprising to anybody with even a passing knowledge of climate change discourse. However,

1. ((carbon|(greenhouse gas)|co2) emissions)
2. ((anthropogenic|manmade|(man made)) global_warming)
3. ((source|emitter|emitters|producers) of greenhouse_gases)
4. ((to (combat|minimize|tackle)) climate change)
5. (((due to)|(caused by)) ((climate change)|(global warming)))
6. ((of global warming) (was|are|is))
7. (in (order|(the (atmosphere|recessions))))
8. (((greenhouse gases)|emissions|gases|(carbon emissions)|pollution) blamed ((for|to) global_warming))
9. ((would|should|to|must) (control|reduce|regulate|regulating|release) greenhouse_gases)
10. (((((global|some|sophisticated|complex|the) climate models)|climate models) (project|suggest|predict)) that)

Figure 9.2 Ten of the Top-Level H-groups That Were Induced Automatically from the NTAP Corpus by Salway and Touileb (2014)

the point is that an automatic technique was able to generate computationally tractable structures that relate to important content in the corpus. The following two case studies will reflect on how such structures were used to investigate climate change discourse. First, I will reflect briefly on the general applicability of the technique.

Because the induction process is unsupervised and does not rely on any linguistic resources, then it should port effectively and cheaply to different language and domains; an exception may be languages with free word order. That said, because the induction process exploits partially overlapping word sequences around key terms, it should be expected to be most effective on large corpora with relatively constrained language use. In other words, it will work best with corpora that consist of a single domain and a single text genre, and especially more stylized varieties of language. Like other unsupervised techniques, the method is sensitive to small changes in the input data—e.g., the order in which snippets are presented. This means that in practice it should be run many times and something should be done to amalgamate the results from the different runs. It should also be noted that the technique is computationally intensive—it took several days to process about 100,000 snippets for one key term. Of course, the speed of computing hardware continues to improve at a fast rate, but the fact that the ADIOS algorithm does not seem to be conducive to parallelization means that at least some big data technologies cannot be exploited. The most important caveat is that we are still working to understand more precisely what information structures the technique does and does not capture. Preliminary results are presented and discussed in the following case studies.

Case Study II: Using Unusually Frequent Information Structures to Highlight Distinctive Content

The 671 induced H-groups mentioned in the previous section were examined by Touileb and Salway (2014) in order to assess how well they highlighted distinctive content within individual climate change blogs; selected results are presented and discussed here. Figure 9.3 presents the top ten H-groups for one of the three blogs that were focused on—*Chimalaya*—ranked according to the RRF metric (Edmundson and Wyllys 1961). RRF (ratio of relative frequencies) is a simple measure of keyness that reflects how much more (or less) something appears in corpus A compared to corpus B, while factoring in the size of the corpora; in this case, the comparison was between each blog in turn and the other blogs. As before, each H-group is presented with brackets and '|'s; additionally, a breakdown of the frequencies of the various forms is given. For example, H-group 2 occurs 1,172 times in total—1,061 times as "developing countries" and 111 times as "poor countries".

1. (impact ((of|for) climate change)): 284 - *impact of climate change (284)*
2. ((developing|poor) countries): 1172 - *developing countries(1061), poor countries (111)*
3. (the (causes|effects) | (consequences|impacts) ((of|for) climate change))): 460 - *the impacts of climate change (224), the effects of climate change (203) , the consequences of climate change (29), the causes of climate change (4)*
4. (climate change (talks|meeting|summit| conference)): 131 - *climate change conference (55), climate change talks (47), climate change summit (21)*
5. ((consequences|impacts) ((of|for) climate change)): 478 - *impacts of climate change (416), consequences of climate change (62)*
6. (\d+ per cent): *695 - \d+ per cent (695)*
7. (tackling climate change): 47 - *tackling climate change (47)*
8. ((to (combat|minimize|tackle)) climate change): 130 - *to tackle climate change (72), to combat climate change (57), to minimize climate change (1)*
9. ((causes|effects) ((of|for) climate change)): 357 - *effects of climate change (345), causes of climate change (12)*
10. (to climate change): 1289 - *to climate change (1289)*

Figure 9.3 The Top Ten H-groups by RRF for the Climate Change Blog *Chimalaya*, Generated by Touileb and Salway (2014)

The idea was that focusing on unusually frequent H-groups in each blog would elucidate the distinctive characteristics of climate change discourse in each blog. Consider how the top ten H-groups for the *Chimalaya* blog highlight characteristics that could be useful for further content analysis. Note that *Chimalaya* is a blog primarily addressing climate-politics issues for the Himalaya region; in the corpus, there are 3,782 posts from this blog, totaling about 3.1 million words.

Many of the H-groups reflect what is already known about the general topical content of the blog, but they also provide a finer-grained view on how the different topics are expressed. The H-groups 1, 3, 5, and 9 indicate *Chimalaya*'s focus on the impacts/effects of climate change rather than its causes: 3 and 9 include both "causes" and "effects," but the frequencies of the different forms show that this blog is much more concerned with the effects. The blog's interests in addressing climate change are highlighted by 7 and 8, with frequent mentions of meetings in 4. Its focus on the kinds of countries that comprise the Himalaya region is indicated by 2.

H-group 2 "((developing|poor) countries)" also highlights an example of framing: The strong preference for the form "developing countries" (f=1061) compared with "poor countries" (f=111) indicates a choice to frame these countries in a positive way. H-group 8 "(to (combat|minimize|tackle)) climate change)" is also interesting for framing analysis. It suggests two different framings on how the climate issue can be addressed. First, there is a rather dispassionate and diplomatic approach—indicated by the form "to minimize climate change." Second, there is a more passionate and confrontational position that is expressed with stronger words—"to combat|tackle climate change." The frequencies of these forms within *Chimalaya* make it clear that this blog is firmly taking the second position (f=1 vs. f=129); this is further supported by H-group 7.

See Touileb and Salway (2014) for more discussion and comparison of the H-groups in *Chimalaya* and other blogs and for a comparison of H-groups with keywords, n-grams, and collocation data. Here, for reasons of space, I will mention just one further interesting example of an unusually frequent H-group from a different blog—"((you|we) (can|should))" from *Its Getting Hot In Here*. This suggests something about the rhetorical strategies used in the blog. The frequencies of its four forms were: "we can" (f=302), "you can" (f=196), "we should" (f=84), "you should" (f=8). The preference for "we" versus "you" suggests that the writers are trying to be inclusive of their readers and are urging for collective action against climate change. The even stronger preference for "can" versus "should" suggests that the writers are trying to maintain an encouraging and positive tone and trying to avoid alienating people by not telling them directly what to do.

Overall, it seems that skimming a list of unusually frequent H-groups can give a researcher a useful impression of the content of a corpus and suggest lines of inquiry—e.g., for analyzing topics, framing, and rhetorical strategies.

Case Study III: Inducing Structures to Analyze International Relations and Climate Positions

As noted previously, the extent to which the local grammar induction technique induces structuring depends in part on how stylized and repetitive the language in the corpus is. When the method was applied to a corpus comprising records of international climate negotiations, which are expected to be rather stylized in form and repetitive in content, then the resulting structures were sufficient to support simple information extraction and the subsequent analysis of country relations and country positions (Salway, Touileb, and Tvinnereim 2014). Structures were induced from a corpus comprising all texts in the Earth Negotiation Bulletin (ENB) volume 12 for the period 1995–2013. All 32,288 sentences mentioning one or more countries were selected, and every mention of a country, or a list of countries, was replaced with the token 'COUNTRY'. This replacement (that used knowledge outside the corpus) was done to make patterning around mentions of countries more explicit. Seven of the resulting 53 H-groups are shown in Figure 9.4.

H-group 1 "(COUNTRY ((supported|opposed) by) COUNTRY)" was used as a regular expression to extract instances where relations between countries were recorded. This gave 1,145 instances of support, and 592 of opposition, often involving multiple countries; recall that 'COUNTRY' may stand for a list of countries. A count was made for each pair of countries in support and opposition, with a distinction made between 'C1 supported by

1. (COUNTRY ((supported|opposed) by) COUNTRY)
2. (COUNTRY (said|noted|recommended|explained|responded|stressed|questioned|addressed|reiterated|reported|urged|amended|invited...)) ***the V-group contains 51 words*
3. (COUNTRY ((clarified|urged|reported) that)
4. (COUNTRY ((presented|demanded|outlined|favored (the|a))
5. (COUNTRY expressed ((disappointment|concern) that)|((support|appreciation) for)|((readiness|willingness) to)|(satisfaction (with the) (outcome|reconstitution|functioning|work) (of the)))
6. (COUNTRY called (((for |on) (parties|(developed countries)) to)|((for a) (cautious|three phased|common|phased|bottom up|budget|global) approach to)|(for an (overview|elaboration|analysis|evaluation|examination) of)))
7. (COUNTRY highlighted ((the (need|basis) for)|(the (benefits|possibility|establishment) of)|(the (consideration|impact|impacts) of)|(the (use|involvement) of)|((the need to) (err|focus) on)|(the (role|importance) (of the))))

Figure 9.4 Seven of the H-groups Automatically Induced from Earth Negotiations Bulletin Texts by Salway, Touileb, and Tvinnereim (2014)

C2' and 'C2 supported by C1', and then a scatterplot was made from these counts in order to visualize relations between countries. This highlighted instances of both strong support (e.g., Canada for the United States) and strong opposition (e.g., the European Union to the Group of 77).

Then H-groups 2, 3, and 4 were combined into a regular expression to extract instances of the statements made by countries. For each country, a file was made with the text following every instance of "COUNTRY said | noted | recommended | (etc.)," until the end of the sentence. The country files were then subject to text scaling using the Wordfish tool (Slapin and Proksch 2008): This places texts on a single induced scale that is intended to mirror political positions. For the 40 countries with the most statements, the values of the parameter indicating country position on the induced scale ranged in ascending order from Austria (-2.38) via Belgium, Germany, the UK . . . New Zealand to Japan (-.62) and then on to Papua New Guinea (-.26), Tuvalu, Peru . . . Iran, Bolivia, Barbados, India, and Algeria (1.44). This result—with a large gap between the countries up to Japan and the countries from Papua New Guinea onward—confirms what is known to be the main cleavage in international climate negotiations between developed and developing countries. Such a comparison of countries would not have been possible by scaling the original ENB texts, because they are organized by date, not country, so there was value in being able to use the induced structures as the basis for extracting each country's statements.

H-groups 1–4 all have a relatively shallow structure. In order to induce further structure, we created new input files, one for each of the most frequent speech acts from 2 to 4—e.g., one file comprised all the sentences containing "COUNTRY expressed." The resulting H-groups, including 5–7 in Figure 9.4, do indeed have richer structures and show in a more nuanced way how countries' positions are reported in the ENB texts. As such, they could perhaps be used to extract more specific information about countries positions, but this is not something we have tried yet.

Closing Remarks

The increasing volume and variety of textual material related to climate change means that it is necessary to use data-driven text analysis techniques as discovery tools in order to elucidate and interpret linguistic patterning. Knowledge of salient linguistic forms is crucial for establishing computationally tractable conceptual frameworks so that the textual realizations of relevant phenomena and concepts can be coded and counted automatically. Taking a data-driven approach to analyzing a text corpus alleviates the problem of "seek and ye shall find," but success depends on the extent to which relevant concepts are realized in regular and repeating ways in texts, as well as on automated text analysis capabilities. While some problems can be tackled by treating texts as bags of words and analyzing language only at the level of lexis, higher-level linguistic patterning must be identified in order to fully realize the potential of automated techniques.

While this can partly be achieved using established corpus linguistic techniques, there is a need for the development of further text mining and visualization tools in order to provide researchers with manageable views of salient linguistic forms in large corpora. Early results from local grammar induction suggest that some meaningful structures can be induced from an unannotated text corpus and that giving a researcher access to these structures is useful in the exploratory phase of an investigation. More work is needed to understand what can and cannot be induced from text alone (i.e., without knowledge of context and general domain knowledge), how to manage the sensitivity of inductive techniques to small change in inputs, and how to incorporate data-driven content analysis techniques into discovery tools with informative visualizations and user interaction, and into methods for automatic coding.

References

Akil, H. Summer 2003. "Scientific strategy in Neuroscience: Discovery Science versus hypothesis-driven research." *Neuroscience Quarterly*: 4–5.

Auer, M. R., Y. Zhang, and P. Lee. 2014. "The potential of microblogs for the study of public perceptions of climate change." *WIREs Climate Change* 5: 291–296.

Baker, P. 2006. *Using corpora in discourse analysis*. London: Continuum.

Baker, P., C. Gabrielatos, M. Khosravinik, M. Krzyżanowski, T. McEnery, and R. Wodak. 2008. "A useful methodological synergy? Combining critical discourse analysis and corpus linguistics to examine discourses of refugees and asylum seekers in the UK press." *Discourse & Society* 19 (3): 273–306.

Baumer, E. P. S., F. Polletta, N. Pierski, C. Celaya, K. Rosenblatt, and G. K. Gay. 2013. "Developing computational supports for frame reflection." *Proceedings of the iConference.*

Boykoff, M. T. 2008. "The cultural politics of climate change discourse in UK tabloids." *Political Geography* 27 (5): 549–569.

Boykoff, M. T., and J. Timmons Roberts. 2007. *Media coverage of climate change: Current trends, strengths, weaknesses*. Human Development Report 2007/2008. UNDP.

Diakopoulos, N., D. Elgesem, A. Salway, A. Zhang, and K. Hofland. 2015. "Compare clouds: visualizing text corpora to compare media frames." *Proceedings of the IUI Workshop on Visual Text Analytics.*

Diakopoulos, N., A. Zhang, D. Elgesem, and A. Salway. 2014. "Identifying and Analyzing Moral Evaluation Frames in Climate Change Blog Discourse." *Proceedings of the International Conference on Weblogs and Social Media (ICWSM 2014).*

Diakopoulos, N., A. Zhang, and A. Salway. 2013. "Visual analytics of media frames in online news and blogs." *IEEE InfoVis Workshop on Text Visualization.*Doulton, H., and K. Brown. 2009. "Ten years to prevent catastrophe? Discourses of climate change and international development in the UK press." *Global Environmental Change* 19 (2): 191–202.

Edmundson, H. P., and R. E. Wyllys. 1961. "Automatic abstracting and indexing—Survey and recommendations." *Communications of the Association for Computer Machinery* 4 (5): 226–234.

Fløttum, K., Ø. Gjerstad, A. M. Gjesdal, N. Koteyko, and A. Salway. 2014. "Representations of the future in English language blogs on climate change." *Global Environmental Change* 29: 213–222.

Grimmer, J., and B. Stewart. 2013. "Text as data: The promise and pitfalls of automatic content analysis methods for political texts." *Political Analysis* 21 (3): 267–297.

Gross, M. 1997. "The construction of local grammars." In: Roche, E., and Y. Schabes (eds.). *Finite-state language processing*, pp. 329–354. Cambridge, MA: MIT Press.

Grundmann, R., and R. Krishnamurthy. 2010. "The discourse of climate change: A corpus-based approach." *Critical Approaches to Discourse Analysis across Disciplines* 4 (2): 125–146.

Grundmann, R., and M. Scott. 2014. "Disputed climate science in the media: Do countries matter?" *Public Understanding of Science* 23 (2): 220–235.

Harris, Z. 1954. "Distributional structure." *Word* 10 (2/3): 146–162.

Harris, Z. 1988. *Language and information*. New York: Columbia University Press.

Hopkins, D. J., and G. King. 2010. "A method of automated nonparametric content analysis for social science." *American Journal of Political Science* 54 (1): 229–247.

Jang, S. M., and P. S. Hart. 2015. "Polarized frames on 'climate change' and 'global warming' across countries and states: Evidence from Twitter big data." *Global Environmental Change* 32: 11–17.

Kirilenko, A. P., T. Molodtsova, and S. O. Stepchenkova. 2015. "People as sensors: Mass media and local temperature influence climate change discussion on Twitter." *Global Environmental Change* 30: 92–100.

Kirilenko, A. P., and S. O. Stepchenkova. 2014. "Public microblogging on climate change: One year of Twitter worldwide." *Global Environmental Change* 26: 171–182.

Koteyko, N. 2010. "Mining the internet for linguistic and social data: An analysis of 'carbon compounds' in web feeds." *Discourse & Society* 21 (6): 655–674.

Mayer, F. W. 2012. "Stories of climate change: Competing narratives, the media, and U.S. public opinion 2001–2010." Joan Shorenstein Center on the Press, Politics and Public Policy Discussion Paper Series #D-72. https://www.hks.harvard.edu/presspol/publications/papers/discussion_papers/d72_mayer.pdf

McEnery, T., and A. Hardie. 2012. *Corpus linguistics*. Cambridge: Cambridge University Press.

Olteanu, A., C. Castillo, N. Diakopoulos, and K. Aberer. 2015. "Comparing events coverage in online news and social media: The case of climate change." *Proceedings of the International AAAI Conference on Web and Social Media (ICWSM 2015).*

O'Neill, S., H. P. Williams, T. Kurz, B. Wiersma, and M. Boykoff. 2015. "Dominant frames in legacy and social media coverage of the IPCC Fifth Assessment Report." *Nature Climate Change* 5: 380–385.

Pearce, W., K. Holmberg, I. Hellsten, and B. Nerlich. 2014. "Climate change on Twitter: Topics, communities and conversations about the 2013 IPCC Working Group 1 Report." *PLoS ONE* 9 (4): e94785. doi: 10.1371/journal.pone.0094785

Salway, A., D. Elgesem, K. Hofland, Ø. Reigem, and L. Steskal. 2016. "Topically-focused blog corpora for multiple languages". *Proceedings of the 10th Web as Corpus Workshop (WAC-X), ACL 2016.*

Salway, A., K. Fløttum, and D. Elgesem. 2015. "Representations of the future in 'accepting' and 'sceptical' climate change blogs." *Proceedings of Corpus Linguistics 2015,* Lancaster University.

Salway, A., K. Hofland, and S. Touileb. 2013. "Applying corpus techniques to climate change blogs." *Proceedings of Corpus Linguistics 2013*, Lancaster University.

Salway, A., and S. Touileb. 2014. "Applying grammar induction to text mining." *Proceedings of the 52nd annual meeting of the Association for Computational Linguistics (ACL 2014)*, 712–717.

Salway, A., S. Touileb, and E. Tvinnereim. 2014. "Inducing information structures for data-driven text analysis." *Proceedings of ACL Workshop on Language Technologies and Computational Social Science.*

Schäfer, M. 2012. "Online communication on climate change and climate politics: A literature review." *WIREs Climate Change* 3: 527–543.

Schäfer, M., and I. Schlichting. 2014. "Media representations of climate change: A meta-analysis of the research field." *Environmental Communication* 8 (2): 142–160.

Schmidt, A., A. Ivanova, and M. Schäfer. 2013. "Media attention for climate change around the world: A comparative analysis of newspaper coverage in 27 countries". *Global Environmental Change* 23: 1233–1248.

Sharman, A. 2014."Mapping the climate sceptical blogosphere." *Global Environmental Change* 26: 159–170.

Shneiderman, B. 2001. "Inventing discovery tools: Combining information visualization with data mining." *Algorithmic Learning Theory* 2225: 58.

Sinclair, J. 2004. *Trust the text: Language, corpus and discourse*. London & New York: Routledge.
Slapin, J., and S.-O. Proksch. 2008. "A scaling model for estimating time-series party positions from texts." *American Journal of Political Science* 52 (3): 705–722.
Solan, Z., D. Horn, E. Ruppin, and S. Edelman. 2005. "Unsupervised learning of natural languages." *Proceedings of the National Academy of Sciences* 102 (33): 11629–11634.
Touileb, S., and A. Salway. 2014. "Constructions: A new unit of analysis for corpus-based discourse analysis." *Proceedings of the 28th Pacific Asia Conference on Language, Information and Computation (PACLIC 28)*.
Tvinnereim, E., and K. Fløttum. 2015. "Explaining topic prevalence in answers to open-ended survey questions about climate change." *Nature Climate Change* 5: 744–747.
Wild, K., A. Church, D. McCarthy, and J. Burgess. 2013. "Quantifying lexical usage: Vocabulary pertaining to ecosystems and the environment." *Corpora* 8 (1): 53–79.

Biographical Notes on Contributors

Dimitrinka Atanasova is a research assistant at Queen Mary University of London. She is interested in health and science communication, framing and discourse analysis, and metaphors. Her work has appeared in journals such as *Environmental Communication*, *Health*, *Public Understanding of Science*, and *Obesity Reviews*. Prior to her PhD, she was a media consultant.

Trine Dahl is professor of English linguistics at the Norwegian School of Economics. She takes a special interest in text linguistic research in the Anglo genre tradition. Being particularly interested in science communication, she has undertaken several studies involving academic texts and knowledge construction in economics. In recent years, her research has been directed at climate change discourse, in the shape of both organizational reports and media texts. She is coauthor (with Kjersti Fløttum and Torodd Kinn) of the book *Academic Voices* (2006). Her work has also been published in journals such as *Journal of Pragmatics*, *Text & Talk*, and *Written Communication*.

Helge Drange is a Norwegian oceanographer and climate scientist, holding a professorship at the Geophysical Institute at the University of Bergen. Drange's research focuses on the development, verification, and use of global-scale general circulation models for climate studies; analysis of naturally occurring variations in climate and longer-term changes caused by human-induced greenhouse gas emissions; and variations and changes in global and regional sea levels. Drange has published in international journals such as *Ocean Modelling*, *Climate Dynamics*, *Quaternary Science Reviews*, and *Geoscientific Model Development*.

Kjersti Fløttum is professor of French linguistics at the Department of Foreign Languages, University of Bergen. Her research fields are text and genre theory and discourse analysis, with a special focus on linguistic polyphony in scientific, political, and climate change discourse as well as on narrative structures in climate change discourse. Fløttum has headed several cross-disciplinary projects (KIAP, EURLING, LINGCLIM). She is coauthor of the books *ScaPoLine* (2004) and *Academic Voices* (2006) and is editor of *Speaking to Europe* (2013). She has published in international journals such as *Journal of Pragmatics*, *Journal of Language and Politics*, *Language & Communication*, *Text & Talk*, *Climatic Change*, *Global Environmental Change*, *Wiley Interdisciplinary Reviews: Climate Change*, and *Nature Climate Change* and in *Oxford Research Encyclopedias*. Together with Anwar Saab, "1001Films," Fløttum is co-producer of the documentary film "Talking about climate," available at https://vimeo.com/178449717.

Øyvind Gjerstad is associate professor of French linguistics at the Department of Foreign Languages, University of Bergen, Norway. Among his main research interests are polyphony and policy narratives, as both linguistic and discursive phenomena. In recent years, he has participated in the LINGCLIM project, headed by professor Kjersti Fløttum, with whom he has coauthored several studies on climate change discourse. Their most recent work is the review article "Narratives in Climate Change Discourse," published in *Wiley Interdisciplinary Reviews: Climate Change* (2016).

Mike Hulme is professor of climate and culture in the Department of Geography at King's College London. His work explores the idea of climate change using historical, cultural, and scientific analyses, seeking to illuminate the numerous ways in which climate change is deployed in public and political discourse. His books include *Can Science Fix Climate Change? A Case against Climate Engineering* (Polity, 2014), *Why We Disagree About Climate Change* (Cambridge, 2009), and *Weathered: Cultures of Climate* (SAGE, 2016). From 2000 to 2007, he was the Founding Director of the Tyndall Centre for Climate Change Research, based at the University of East Anglia. And since 2007, he has been the founding Editor-in-Chief of the journal *Wiley Interdisciplinary Reviews (WIREs) Climate Change*.

Michael D. Jones is an assistant professor at Oregon State's School of Public Policy. His research interests include the policy process, narrative and public policy, and science communication. Some of his more recent research has appeared in journals such as *Critical Policy Studies*, the *Policy Studies Journal*, and *Social Science Quarterly*. He is also coeditor (along with Elizabeth Shanahan and Mark K. McBeth) of the edited volume *The Science of Stories: Applications of the Narrative Policy Framework in Public Policy Analysis* (2014).

Oleksandr Kapranov is a postdoctoral fellow at the Department of Foreign Languages, University of Bergen. His research interests include cognitive linguistics and psycholinguistics, with a particular focus on conceptual metaphors, elucidated in a variety of settings, involving metaphors associated with the issue of global climate change. Kapranov's recent research papers have been published in *Topics in Linguistics*, *Lodz Papers in Pragmatics*, and *Text and Discourse*.

Nelya Koteyko is a Reader in Applied Linguistics at Queen Mary, University of London. Her research lies at the intersection of linguistics and social studies of science, medicine, and technology; she has published in these areas in international journals including *Applied Linguistics*, *Science Communication*, *Discourse and Society*, *Global Environmental Change*, and *WIREs Climate Change* and co-edited the volume *Climate Change Communication and the Internet: Challenges and Opportunities for Research* (Routledge, 2017).

Andrew Salway is a senior researcher at Uni Research in Bergen, where he leads the Language and Language Technology Group and works with the Center for Big Data Analysis. He has some 20 years of experience of developing automatic text analysis techniques and incorporating them into humanities and social science research methods, with particular interests in narrative and multimodality, and latterly social media. Recent collaborations have led to publications in *Style* and *Global Environmental Change* and to presentations at conferences in corpus linguistics, social media analytics, digital humanities, and natural language processing.

Index

For Product Safety Concerns and Information please contact our EU
representative GPSR@taylorandfrancis.com
Taylor & Francis Verlag GmbH, Kaufingerstraße 24, 80331 München, Germany

www.ingramcontent.com/pod-product-compliance
Ingram Content Group UK Ltd.
Pitfield, Milton Keynes, MK11 3LW, UK
UKHW022120190926
475702UK00013B/1549
9780367365820